TRADE POLICY, PROCESSING AND NEW ZEALAND FORESTRY

This book examines core issues with respect to the effect of export restrictions; the impact on processing and welfare; the consequences of foreign ownership of the resource; and the possibility of utilising export restrictions as a retaliatory strategy against escalating tariff structures. It also examines the impact of liberalisation of processed good markets. The book employs a combination of formal general equilibrium modelling and counterfactual simulation using computable general equilibrium (CGE) techniques, with the New Zealand forestry industry used as a case study throughout. The book makes a contribution to the literature in this field by incorporating foreign ownership into an extensive formal analysis of processing incentives, developing a new CGE model of the New Zealand economy, utilising this model to evaluate the costs of export restrictions, and utilising the GTAP to provide insights into the possible effect of the APEC Early Voluntary Sector Liberalisation strategy.

Trade Policy, Processing and New Zealand Forestry

JOHN GILBERT
Department of Agricultural Economics
Washington State University

Ashgate
Aldershot • Burlington USA • Singapore • Sydney

Published by
Ashgate Publishing Limited
Gower House
Croft Road
Aldershot
Hampshire GU11 3HR
England

Ashgate Publishing Company
131 Main Street
Burlington
Vermont 05401
USA

Ashgate website: http://www.ashgate.com

British Library Cataloguing in Publication Data
Gilbert, John
Trade policy, processing and New Zealand forestry. - (New voices in economics)
1. Forests and forestry - Economic aspects - New Zealand
2. Forest products industry - New Zealand 3. Forest products industry - Government policy - New Zealand 4. New Zealand - Commerce - Government policy
I. Title
338.1'749'0993

Library of Congress Catalog Card Number: 99-85923

ISBN 0 7546 1064 0

Printed and bound by Athenaeum Press, Ltd.,
Gateshead, Tyne & Wear.

Contents

List of Figures

List of Tables

Preface

I am pleased that John Gilbert asked me to write a few words for the preface of this book because of my warm memories about how the enterprise started and my delight at how it has finished.

In 1992 I dropped by the University of Auckland economics department to give a seminar and visit my friends Geoff Braae and Ken Jackson. This resulted in short visits over the next five New Zealand winters to teach applied general equilibrium modelling at Auckland. John was a student in the first course I taught there. Subsequently, he was my teaching assistant, and briefly my colleague at Duke.

On one of those winter evenings, while I was drinking a New Zealand chardonnay with Gill and Andy Brewis on the rug in front of their logfire, Andy complained that the wooden reservoirs that his company fabricated in Auckland and shipped to the Pacific Islands were getting awfully expensive to produce. This he explained was because the Japanese were buying cheap New Zealand logs and fabricating them into final goods in Japan, putting him at a disadvantage. His solution was to impose an export tax on logs, designed to balance the Japanese import tariffs on manufactured wood products, and thereby redress the advantage that the Japanese trading regime gives to its own producers of manufactured wood products and give his firm a level field to play on.

John and I discussed the issue and realised that the openness of the New Zealand economy in recent years to foreign investment had resulted in foreign ownership of much of New Zealand's forest land. Thus, it seemed that if one were to make a case for taxing exports of logs, it would have to hinge on the premise that such a tax would redistribute rents from foreign owners of the New Zealand forest to New Zealand fabricators of wood products.

I was delighted when John produced this book thoroughly exploring that issue, among others. Here we have a book that takes important real world problems, and applies state of the art theory, modelling techniques (including the GTAP model) and software to explore alternative potential solutions. The book is exceptionally good work. It is extremely well organised and well expressed. It does a fine job of performing calculations that sort out the welfare consequences and implications for income distribution of various tax-subsidy schemes for dealing with the New Zealand forestry sector. The applications are tied in nicely with the appropriate theory of international trade and common sense economics.

The fact that taxes and restrictions on exports of unprocessed forest products and other raw materials are so prevalent throughout the world (existing as they do for example in Canada, Indonesia and the US), make this an important and topical piece of analysis. The author has done a handsome job with the project. The book will appeal to those in the industry, regulators of it, and teachers of forestry economics and trade theory.

Edward Tower
Professor of Economics, Duke University

Acknowledgements

This book is a revision of a Ph.D thesis submitted to the Department of Economics at the University of Auckland. I would like to thank my supervisors, Dr. Ravindra Ratnayake and Mr. Robert Scollay, who provided me with guidance and encouragement over the course of my work on this topic. I would also like to thank the thesis examiners, Professor Peter Dixon and Professor Edward Tower, for comments that have been very helpful in writing this revision. Professor Ippei Yamazawa also provided me with very helpful comments during the early stages of my work.

For financial support provided at various stages during the completion of this work I would like to thank the Japanese Ministry of Education (Monbusho), the APEC Study Centres of Hitotsubashi University and New Zealand, and the Sasakawa Foundation. I would also like to thank the GAMS Corporation for kindly providing me with some of the software necessary to complete my research, and the Department of Economics at the University of Auckland for giving me access to excellent research facilities.

Finally, I would like to thank my wife Hisako, my friends and fellow Ph.D students, and my family for their help and moral support.

List of Abbreviations

AES	Allen (Uzawa) Partial Elasticity of Substitution
APEC	Asia Pacific Economic Cooperation
ASEAN	Association of South East Asian Nations
CES	Constant Elasticity of Substitution
CGE	Computable General Equilibrium
CGTM	CINTRAFOR Global Trade Model
CNIPS	Central North Island Planning Study
CRESH	Constant Ratio Elasticity of Substitution Homothetic
CV	Compensating Variation
DC	Developed Country
DES	Direct Partial Elasticity of Substitution
EEC	European Economic Community
EMDTI	Export Market Development Taxation Incentive
EMIA	Export Manufacturing Investment Allowance
EV	Equivalent Variation
EVSL	Early Voluntary Sector Liberalisation
FAO	United Nations Food and Agriculture Organisation
FCCC	Framework Convention on Climate Change
GAMS	General Algebraic Modelling System
GATT	General Agreement on Tariffs and Trade
GDP	Gross Domestic Product
GEMPACK	General Equilibrium Modelling Package
GNP	Gross National Product
GTAP	Global Trade Analysis Project
GTM	Global Trade Model
HOS	Heckscher-Ohlin-Samuelson
IETI	Increased Exports Taxation Incentive
IO	Input-Output
IZEF	Iterative Zellner-efficient
LDC	Less Developed Country
LP	Linear Programming

MFN	Most-Favoured Nation
NAFTA	North American Free Trade Area
NDI	Net Domestic Income
NZOSG	New Zealand Owned Sawmillers' Group
NZSIC	New Zealand System of Industrial Classification
OECD	Organisation for Economic Cooperation and Development
OLS	Ordinary Least Squares
OPEC	Organisation of Petroleum Exporting Countries
ORANI	Model of the Australian Economy (Dixon et al.,1982)
PEP	Project on Economic Planning (Victoria University of Wellington)
PNW	Pacific Northwest (United States)
ROW	Rest of World
RWE	Roundwood Equivalent
SES	Shadow Elasticity of Substitution
SNA	(New Zealand) System of National Accounts
SUR	Seemingly Unrelated Regressions
TAMM	Timber Assessment Market Model
UNIDO	United Nations Industrial Development Organisation
WAMM	World Assessment Market Model
WTO	World Trade Organisation

1 Introduction

In many countries that depend heavily on the export of various forms of raw materials we can find policies designed to restrict those exports, in order to develop processing capabilities. Where those policies do not exist, we can invariably find calls for their institution. There is a common perception, even amongst some economists, that by exporting raw materials these countries are somehow not performing to their full potential. In New Zealand, much of the debate surrounding the use of processing incentives and restrictions of raw material exports has been focused on the forestry industry.

The New Zealand forestry sector is amongst the oldest and most historically, politically, and economically significant sectors in this country's European history. Recognition of the importance of New Zealand's forest reserves and their potential for economic exploitation dates back to at least 1769, when it was observed by Cook that the forest trees of New Zealand would, "with suitable preparation, be such masts as no country in Europe could produce" (cited in Roche, 1990, p.14). New Zealand was in fact used as a source of spars for the British Admiralty decades before the Treaty of Waitangi was signed in 1840, and a domestic timber industry to supply the demands of colonisation emerged not long after. These demands eventually took their toll on the indigenous forest resource, leading to the creation of exotic plantations at the impetus of the state, in an effort to ensure a steady domestic supply. In more recent years, in particular those following the major restructuring of the industry that took place in the latter half of the 1980s, the forestry industry has, perhaps ironically, come under increasing scrutiny for its potential as a major source of export revenue for New Zealand, as supplies from plantations have begun to vastly outstrip demand. The sector is also a significant employer, and contributor to national income. Moreover, it is an industry that has come to embody the current New Zealand debate on foreign ownership, Maori land rights, environmental concerns, and the costs and benefits of encouraging the domestic processing of New Zealand's resources.

However, despite its substantial importance to the New Zealand economy, surprisingly little extensive research has been undertaken on the forestry sector. This is particularly so with respect to trade policy aspects. Indeed, the vast majority of economic modelling effort within the forestry sector has been devoted to estimating the supply of wood, and the economics of different management regimes. It is a well accepted fact that there has been an imbalance of effort within New Zealand which has led to numerous models of the forest resource and limited attention to the market, and in particular to models of the small open economy and international trade.[1] This is surprising, since there are a number of significant areas of debate in the trade policy area (in particular relating to the issue of further processing), and a number of other issues that, while not the subject of major contention, have yet to be empirically examined (such as the benefits of liberalisation in export markets). It is the intention of this study to go some way towards correcting this imbalance of effort.

The study identifies and examines three core issues with respect to the issue of log export restrictions; the impact on processing and welfare; the effect of foreign ownership of the resource; and the possibility of utilising export restrictions as a retaliatory strategy against escalating tariff structures. It also examines the impact of liberalisation of forestry products trade on a region-wide basis.

The first issue is a long-standing one. New Zealand has a tendency to export largely unprocessed forest products (i.e., logs) rather than add value to the product within New Zealand, a tendency that has become increasingly prominent over recent years. While New Zealand has been taking a laissez-faire approach to the industry, this is not the case in other countries. For example, Indonesia has been utilising export restrictions on unprocessed roundwood to encourage domestic processing for a significant period of time now.[2] Moreover, Indonesia is not an isolated example, virtually all the wood producing economies in the Asia-Pacific region have imposed restrictions of some kind on the exporting of roundwood (FAO, 1993, p.156). Export restrictions have also been used in New Zealand at various times in the past, and while calls for export restrictions are still made from time to time in New Zealand (usually by sawmillers concerned at paying world prices), it is perhaps surprising how little attention has been paid to the economic analysis of the issues that arise from such a policy proposal. One can find numerous government studies and papers replete with references to the desirability of increasing the level of domestic processing, but little in the way of concrete analysis. There are legitimate questions to be answered with respect to the benefits of policy induced increased domestic processing. Is it possible to increase welfare at the same time as increasing processing? Is restricting exports an appropriate policy intervention to increase the level of processing, or are other policy interventions likely to be less damaging in welfare terms?

A second issue, which makes the general analysis of export restrictions more complex, is the addition of another of the key focal points of more recent debate: the extent to which foreign ownership of the forestry resource has come about due to the government program for the sale of state forestry assets, which began in 1989. Currently the level of 100 percent foreign owned forest assets is around 18 percent of the total resource (nearly 50 percent if foreign shareholdings of the major New Zealand forestry companies - Carter Holt Harvey and Fletcher Challenge - are included). While much of the debate surrounding the issue of foreign ownership has focused on the issue of sovereignty, an issue difficult to deal with from an economic perspective, there is also the question of to what extent it is desirable to have foreign ownership of the forest resource. Much of the political rhetoric has focused on the supposed job creation and growth benefits of foreign investment, and the reduction in debt that the sale proceeds allow. However, when the investment concerned involves a mere transfer of existing resources from domestic to foreign control it is unclear as to from where these benefits arise. While these issues are largely beyond the scope of this study, the issue also poses an interesting question with respect to the arguments for export restrictions. Given foreign ownership of the forest resource, a log export tax may effect a transfer of income from foreign to domestic sources. How might this affect optimal strategies?

The third issue identified above relates to the tariff structures of the major wood importing countries in the region. Almost all have escalating structures (Japan and Korea in particular), designed to encourage importing raw materials for processing within the importing country. The existence of escalating tariffs may penalise New Zealand processing. However, the question as to whether retaliation by imposing export restrictions (which directly hurt the overseas processing industries in importing countries that rely on imported logs) could be an appropriate strategy, and under what circumstances, remains unanswered. Moreover, little is known about the extent of the effect of escalating tariffs, or the likely impact of their removal under the auspices of the APEC liberalisation agenda.

These issues are not unique to forestry, nor to New Zealand. However, none have been formally examined or empirically quantified in the New Zealand context. Moreover, the general literature, in particular that utilising general equilibrium methodology, is also relatively scarce. The objectives of this study are therefore as follows:

a) To analyse in a formal general equilibrium framework that accounts for important intra-sectoral, inter-sectoral and regional linkages, the impact of export restrictions and other processing incentives, and of liberalisation in other markets, on processing, welfare, and income distribution.

b) To use these formal results as the foundation of empirical study with the objective of determining the direction and quantifying the extent of changes in relevant economic variables, in particular economic welfare, as a result of export restrictions and other processing incentives for New Zealand forestry.
c) To use these formal and empirical results to provide insights into the policy debate over export restrictions and processing in New Zealand, with reference to those issues identified above.

The perspective of this study is that of international trade theory, rather than forestry economics, and the methodology employed is a combination of formal economic modelling, and counterfactual simulation using computable general equilibrium (CGE) modelling techniques. Whilst CGE techniques have a number of well-known disadvantages (discussed in detail in Chapter 6), they also have numerous advantages. In particular, the models used here are based on clearly defined and described structural models, there is a high degree of theoretical and logical consistency, and we are able to incorporate facets of the processing issue that can only be done in a general equilibrium framework. We therefore believe that CGE is the best available technique for dealing with the issues identified in this study.

This study makes the following contributions to the existing literature on processing incentives and trade:

a) Incorporating the issue of foreign ownership of natural resources into an extensive formal analysis of processing incentives, which in addition treats the forests as an intermediate good (in line with New Zealand's emphasis on plantation forestry) rather than as an endowment.
b) Developing a new, trade focused CGE model of the New Zealand economy with an emphasis on the forestry and forest processing sectors, which incorporates econometric estimates of key relational parameters.
c) Utilising the above model to provide estimates of the effects of export restrictions and other processing incentives under a variety of assumptions on the New Zealand economy, and thus making possible a more informed policy debate.
d) Utilising an existing global trade model (the GTAP) to provide insight into the possible effect of the removal of barriers to forest products trade in other economies, and in particular information on the possible effects of the APEC Early Voluntary Sector Liberalisation (EVSL) program.

The study is divided into three main sections. The first is largely background material, and is introductory in nature. Chapter 2 contains an overview of the New Zealand forestry sector and its place in the New Zealand and global economies, to familiarise those who are new to the sector. Readers familiar with the features of New Zealand forestry may prefer to skip this chapter and delve directly into Chapter 3, which contains a description of current issues facing the sector, in a sense defining more clearly the motivation for this study, and a review of some of the arguments for processing incentives. This chapter also contains some discussion of the international agreements covering the use of such incentives. Finally, Chapter 4 is a review of the existing applied literature on this topic.

The second section of the study, Chapter 5, takes each of the three aspects of processing incentives discussed above and analyses them in turn in a formal mathematical framework. Here export restrictions and other processing incentives are examined in the context of a number of formal general equilibrium models, beginning with the price-exogeneous (small country) case, and turning later to a two-country trading equilibrium model. The purpose of this chapter is twofold. First is to formally describe the impact of intervention on factor prices, output, trade and utility. The models presented in this chapter therefore serve to clarify and formalise the underlying structure of the larger computable models that follow. Second is to identify those aspects of the problems that are not amenable to a neat theoretical solution, and for which other methods must be employed. This chapter may be useful for those interested in theoretical aspects of processing incentives, regardless of the industry involved.

The final section of the study is the largest, and concerns the empirical application of the theoretical models developed in Chapter 5 to the New Zealand context. Chapter 6 describes a computable general equilibrium (CGE) model that we develop of the New Zealand economy, which is specifically designed with a focus on the forestry sector. This chapter also contains details of the data used in the study, in particular base year equilibrium data, data for parameter estimation, and external sources of parameters. The chapter intentionally contains a considerable amount of technical detail on the structure of the model, the data, and the techniques of implementation. Again, it is hoped that this material may be useful for those considering similar exercises for related problems, but much of the material can be safely skipped for those who are not interested in the finer points of the modelling. The model developed is used for detailed analysis of the export restriction debate in Chapter 7, which contains the results of simulating various policies within this framework, and discussion of their policy implications. We are able to use the model not only to give us unambiguous answers where a theoretical model cannot, but also to estimate the likely magnitude of effects on welfare, factor incomes, outputs, resource pulls, etc. Chapter 8 considers the issues detailed above by making use of a major recent development in the CGE literature, the GTAP. The GTAP is a global trade model, incorporating data on all of New Zealand's major trading partners. Hence, although its use raises some compatibility issues, the GTAP is particularly suitable for analysing the effect of liberalisation in forestry by other countries. In addition, since the GTAP allows us to account for inter-regional linkages in a manner not possible with a single country model, it also serves a role in helping to understand the sensitivity of the results of Chapter 7 to the single country model specification. Finally, Chapter 9 contains a summary and conclusions.

Notes

[1] See Leslie (1986), Hunter (1989) and Whyte (1989), and discussion in Chapter 3.

[2] The Indonesian example is discussed at further length in Chapter 3. See also Wijewardana (1989) for an overview, and Lindsay (1989) and Sidabutar (1988) for examples of studies attempting to estimate the cost of export restrictions for Indonesia.

2 An Overview of the New Zealand Forestry Sector

Introduction

In this chapter we examine the current state of, and trends evident in, the New Zealand forestry sector. The objective of this chapter is to provide a brief summary of the scope and structure of the sector, and to highlight some of the more important aspects of the sector's structure, performance, and contribution to the New Zealand economy. As this is a study of economic policy issues, it is not possible to provide an in depth history of the sector, although the topic is of considerable interest.[1] Nor is it possible to provide details on the processing or marketing of forest products. Other works written by forestry experts dealing with these issues are available. Other useful overviews are provided by Brown (1997) and Le Heron (1997).

The chapter is divided into three sections. The first is concerned with domestic aspects of the New Zealand forestry sector, and covers the structure of the forestry sector and its place in the New Zealand economy. The sector is described in terms of the size of the forestry resource, and its contribution to the economy in terms of employment, and GDP.

The second deals with export trends. Figures for overall export growth in both value and absolute volume terms are provided for analysis. Export destinations are also examined.

The third views the New Zealand forestry sector from an international perspective. Data detailing world forestry production is discussed. New Zealand's position in the international arena is briefly commented on. A summary and concluding comments are presented at the end of the chapter.

Domestic Aspects

The New Zealand Forestry Resource

New Zealand's forestry resource can be classified into two categories, those resources coming from planted production forests, and those coming from natural production forests. Brown (1997) refers to the separation of natural and plantation resources as a principle of 'dual' forest estates. These estates are distinct in terms of their legal, institutional and functional dimensions.

The plantations are among the largest exotic forestry plantations in the world, and consist mainly (approximately 89 percent) of radiata pine, a softwood species introduced and developed by the state as a fast growing replacement for rapidly declining stocks of natural timber earlier this century. Approximately 5 percent of the plantations are Douglas Fir, and 2 percent hardwoods. Special purpose plantation species have also been evaluated, and are increasingly being planted by small scale forest growers, the most popular varieties being macrocarpa and blackwood. The plantations are largely commercial in nature, with some overseeing but minimal intervention by the Ministry of Forestry. There is general acceptance that the growing of plantations for harvest, by substituting for natural forest exploitation, is a method of natural forest conservation.

Those resources that are classified in the natural forest category are in general managed by the New Zealand Department of Conservation, with soil and water conservation and recreational values as the primary objectives. These forests can be divided into two main types, the beech forests, dominated by one or more of the four indigenous species of nothofagus, and the conifer-hardwood forests dominated mainly by podocarps (Brown, 1997). Severe restrictions are placed on the utilisation of natural resources, including those that have been set aside for production, by both the Resource Management Act 1991 and the Forest Amendments Act 1993, including restrictions of the annual cut and a strict insistence on full utilisation. Those resources that fall into the planted production forests category are or will be available for economic exploitation, and are the key focus of this study.

Table 2.1 gives a relative description of the size of the New Zealand forestry resource. Forests occupy approximately 29 percent of New Zealand's total land area. Of this area, approximately 18 percent, or 1.4 million hectares (roughly 5 percent of New Zealand's total land area), is in the plantations. Much of the planted resource is, however, still immature, the weighted average age for all New Zealand being 13 years in 1991, while the average crop rotation period is approximately 30 years (Ministry of Forestry, 1992). The vast majority of New Zealand's land is used by the agricultural sector (approximately 50 percent), with the balance being made up of urban land, mountains, scrublands and other unusable areas.

Ownership of the Resource

Of the natural forestry resource, some 79 percent is held by the state in the form of national parks, scenic reserves, etc., the remainder being in freehold or leasehold land. With respect to the plantations, although many of the earlier plantation forests were developed by the state, development initiative has come largely from the private sector over the past twenty years. The state has not planted any new land since 1989. While prior to the sales of state forestry assets ownership of plantation forests was shared almost evenly between the public and private sectors, at present approximately 80 percent of the plantation resource is in private hands.

Table 2.1: Areas of New Zealand Forested and Non-Forested Land

Type of Land Cover	**Estimated Area (000 hectares)**		**Percentage of Total NZ Land Area**	
Natural Forest	6406		23.0	
Plantation Forest	1388		4.8	
Total Forested Land		**7794**		**28.8**
Grassland and Lucerne	9600		35.0	
Fruit, Vegetables and Nurseries	104		0.3	
Crops	309		1.2	
Tussock and Danthonia	3917		14.9	
Other Non-Forested Land	5246		19.4	
Minor Offshore Islands	83		0.3	
Total Non-Forested Land		**19259**		**71.2**
Total New Zealand Land Area		**27053**		**100.0**

Source: Ministry of Forestry (1996)

Foreign ownership of the forest resource has recently come about due to the sale of management and cutting rights for state owned commercial forestry assets, which was launched by the New Zealand Forestry Corporation Limited on 25 October 1989. The philosophy underlying the sale was a belief that private owners would utilise the resource more efficiently and profitably than the government. This was not a new notion by any means, and can be traced back to Schmitt (1972) with respect to the New Zealand forestry sector. Rights to approximately 550 thousand hectares of plantation forests were made available for sale. In the sale process, the government made no distinction between domestic and overseas bidders. The inclusion of foreign bidders was to ensure maximum competition in the process, and thereby maximise the returns accruing to the government.

The sale did not include the land itself (i.e., only the wood resource and the right to plant and harvest further was sold) or Maori leased forests, and the government kept the right to remain the owner of the forests if the bids received were unacceptable. An agreement between the Crown, the Maori Council and the Federation of Maori Authorities was made to provide security of tenure for purchasers of state plantations and to protect the interests of Maori who have claims before the Waitangi Tribunal.[2] The system works such that purchasers have the right to use the land for a period sufficient to permit any existing tree crop to reach maturity and be harvested. The right to use the land is automatically extended by one year each year unless notice of termination is given. If notice of termination is given the purchaser retains the right to harvest any tree crops planted prior to this notice. The state pledges to compensate any successful claimants to the Waitangi Tribunal for the notice period.

Management and cutting rights to 249 thousand hectares were initially allocated to private owners, while the rights to 305 thousand hectares remained in the hands of the state. The major purchasers in the first sales round were the established New Zealand forestry companies Carter Holt Harvey Limited and Fletcher Challenge, but several overseas based companies also made significant investments. The largest single foreign purchaser at this stage was a Japanese company, Juken Nissho Limited, which purchased 43 thousand hectares.

Where management and cutting rights were not sold, plantations were initially transferred to three new state owned enterprises: Forestry Corporation of New Zealand Limited, Timberlands West Coast Limited and New Zealand Timberlands Limited. New Zealand Timberlands Limited was subsequently sold, thereby transferring a further 97 thousand hectares into private ownership. The remaining forests continued to be managed for the state. In 1996, the government completed the sale of state forestry assets, with the sale of the remaining forests at Kaingaroa to a consortium led by Fletcher Challenge.[3]

At present, approximately 18 percent of New Zealand's plantation resource is directly owned by foreign interests. This only includes 100 percent ownership of cutting rights, however. Total foreign interests (taking into account foreign stakes in the large New Zealand forestry companies) in the industry are significantly higher. Roche (1992) estimated the figure at 36 percent. Currently, if we include Carter Holt Harvey as foreign owned (51 percent of this company was purchased by American-based International Paper Limited in 1994, making it effectively foreign-controlled) and the 37.5 percent of the latest sale to the Fletcher Challenge Consortium that is now owned by the China International Trust (in turn owned by the Chinese Government) we get a figure in the region of 48 percent, a not insubstantial proportion.[4] Some issues relating to the increase in foreign ownership of the forest resource are discussed in further detail in the following chapters.

Industrial Structure and Production

The forestry sector can be broadly split into two component industries or sub-sectors, which are referred to as the timber industry and the forest products industry. The timber industry consists of the production of roundwood, the production of sawn timber, the production of wood chips, and timber preservation. The forest products industry is characterised by a higher level of value-added production, and consists of the pulp and paper industry and the wood-based panels industry.

The roundwood sub-industry involves the removal for either further processing or export of the forestry resource. In 1995, an estimated 16.4 million cubic metres of roundwood were removed from New Zealand forests. Some 16.2 million of these removals, or 99 percent, was from plantation forests. The output is used to support the sawn timber sub-industry, which produced 2.9 million cubic metres of sawn timber in 1995, approximately 97 percent of which was from exotic plantation species. The wood chip sub-industry uses both native and exotic trees unsuitable for sawn timber production, in addition to forest and sawmill residues.

The pulp and paper sub-industry, concentrated in the North Island of New Zealand, takes approximately 28 percent of total roundwood removals as fibre input, while the wood based panels sub-industry takes approximately 1.4 percent in addition to waste residues. Total production of pulp reached just over 637 thousand air dry tonnes in 1995. Paper and paperboard production reached 876 thousand tonnes over the same period, while total production of veneer, plywood, particle board, and fibreboard reached 1.3 million cubic metres.

Table 2.2 provides estimated four-firm concentration ratios for forestry related industries (data is only available for manufacturing industries, with the exception of chipmills). The four-firm concentration ratio is the sum of the market shares of the four largest firms in a given industry, and is a common indicator of the degree of competitiveness, with higher ratios indicating less competitiveness. As can be seen, the concentration ratios of pulp and paper production, and production of plywood are somewhat more concentrated than the New Zealand average, while sawmilling is less concentrated. This is presumably a reflection of the larger investment required for the higher value-added processing industries, since large investment requirements may be a form of barrier to entry, which thereby increases concentration.

Table 2.2: Four-Firm Concentration Ratios Based on Turnover Adjusted for Imports and Exports (1987)

Industry	Concentration Ratio
Plywood, veneer and board	84
Pulp, paper and paperboard	79
Planing, preserving and seasoning timber	51
Sawmills	38
NZ Industry Average	49

Source: Statistics New Zealand (1991)

Contribution to the New Zealand Economy

Table 2.3 shows the value added contribution to gross domestic product by production group for New Zealand for the period 1985 to 1994 (in nominal dollars). Together forestry and logging, manufacture of wood products, and manufacture of paper products contributed 4.63 billion dollars to GDP in 1994. This amounts to over 5.5 percent of GDP. Comparing this to the proportion in 1990 of 4.5 percent (in 1985 the figure was 5.2 percent) highlights not only the substantial contribution that the forestry sector makes to the New Zealand economy, but also the rapid growth the sector has experienced over the last few years.

A notable aspect is the substantial increase in the contribution to GDP from forestry and logging, coupled with much smaller changes in the contributions from the two manufacturing sectors. It has frequently been argued that the recent improvements in the performance of the forestry sector have come about largely by increasing exports of unprocessed logs, rather than through attempting to add value within New Zealand. These figures tend to confirm this view. This issue forms a core part of this study, and is discussed at length in the following chapter.

Although the days of the forestry industry being used by the state as a form of social welfare by providing jobs for the unemployed are now over, the forestry sector remains a significant employer. According to Ministry of Forestry (1996), as at mid February 1995, 25,415 people were engaged in forestry and first stage processing, up from 18,239 at the same time in 1990. Table 2.4 gives a complete breakdown of the figures over the period 1990 to 1995. The largest component of the industry in terms of employment is the forestry and logging industry itself (accounting for approximately one third

Table 2.3: Contribution to Gross Domestic Product (GDP) 1985-1994 (Values in Nominal $NZ million)

Production Group	**1985**		**1990**		**1994**	
	Value	**%**	**Value**	**%**	**Value**	**%**
Agriculture	3138	7.98	4491	6.35	4973	6.19
Fishing & Hunting	130	0.33	249	0.35	263	0.33
Forestry & Logging	300	0.76	615	0.87	1437	1.79
Mining & Quarrying	422	1.07	836	1.18	1121	1.40
Food & Beverages	2650	6.74	4015	5.68	4526	5.64
Textiles & Apparel	763	1.94	864	1.22	825	1.03
Wood & Wood Products	621	1.58	752	1.06	1030	1.28
Paper & Printing	1124	2.86	1827	2.58	1982	2.47
Chemicals, Petroleum	998	2.54	1776	2.51	1795	2.24
Non-Metallic Minerals	402	1.02	413	0.58	491	0.61
Basic Metal Industries	359	0.91	460	0.65	645	0.80
Fabricated Metal	2179	5.54	2840	4.01	3050	3.80
Other Manufacturing	102	0.26	151	0.21	156	0.19
Electricity, Gas & Water	1116	2.84	2134	3.02	2228	2.77
Construction	2222	5.65	3216	4.55	2289	2.85
Trade & Restaurants	7274	18.49	10292	14.55	12675	15.79
Transport & Storage	2143	5.45	3687	5.21	3925	4.89
Communications	1034	2.63	2183	3.09	2508	3.12
Financing, Insurance, etc.	4817	12.24	10740	15.18	11517	14.34
Owner-Occupied Dwellings	1929	4.90	5275	7.46	6274	7.81
Social Services	1486	3.78	2720	3.84	3705	4.61
Bank Service Charge	-1285	-3.27	-2996	-4.24	-2655	-3.31
All Market Groups	**33922**	**86.21**	**56539**	**79.92**	**64761**	**80.65**
Non-Market Groups	**4739**	**12.04**	**9112**	**12.88**	**9666**	**12.04**
All Production Groups	**38661**	**98.26**	**65651**	**92.80**	**74427**	**92.69**
Import Duties	604	1.54	604	0.85	616	0.77
Other Indirect Taxes	81	0.21	87	0.12	93	0.12
Goods & Services Tax	-	-	4400	6.22	5161	6.43
Total	**39346**	**100.00**	**70742**	**100.00**	**80297**	**100.00**

Source: Statistics New Zealand INFOS, Contribution to GDP by Industry (Annual)

of the total employment in the sector). In terms of the New Zealand economy as a whole, approximately 1.5 per cent of the total labour force is employed in forestry and first stage processing.

Table 2.4: Employment in the New Zealand Forestry Industry 1990-1995 (Persons Engaged[a])

Activity	**1990**		**1995**	
	Persons	**%**	**Persons**	**%**
Forestry and Logging	5881	0.37	9912	0.57
Sawmills	5199	0.33	6159	0.36
Planing & Preserving	724	0.05	1585	0.09
Chipmills	64	0.00	39	0.00
Plywood, Veneer etc	1152	0.07	2766	0.16
Pulp and Paper	4396	0.28	3629	0.21
Logging Haulage	823	0.05	1325	0.08
Forestry & 1st Stage Processing	**18239**	**1.14**	**25415**	**1.47**
Total Labour Force[b]	**1596100**	**100.00**	**1728000**	**100.00**

[a] Persons engaged is the total number of full-time employees and working proprietors plus half the part-time employees and working proprietors as at 15 February.

[b] Total labour force as at March quarter (Household Labour Force Survey).

Source: Ministry of Forestry (1996)

Demand and Supply Trends

Table 2.5 gives the Ministry of Forestry's regional base cut wood supply forecasts until the year 2025. The available wood supply from plantation forests is expected to grow rapidly over the next thirty years, with large supply volumes coming on line from 2000 (often referred to as the encroaching 'wall of wood'), with a brief pause between 2011 and 2015.

With respect to demand trends, Table 2.6 shows the estimated New Zealand consumption of forestry products in roundwood equivalents for the years 1981 to 1995.[5] Domestic consumption has been characterised by fluctuating cycles of between 4.5 and 6.8 million cubic metres per annum, the main cause being the boom/bust nature of the building industry. Considering the New Zealand market until the turn of the century, Maughan

Table 2.5: New Zealand Regional Base Cut Wood Supply Forecasts ($000m^3$ per year)

Region	2001-05	2006-10	2011-15	2016-20	2021-25
Northland	1756	3211	4303	4367	4692
Auckland	1530	1649	1654	1640	1985
Central North Island	11402	11737	12079	12272	12677
East Coast	1306	1885	2165	2547	3995
Hawkes Bay	1903	1943	2006	2018	3254
Southern North Island	1730	1837	1843	2282	4017
Nelson/Marlborough	2538	2792	2924	2932	3268
Canterbury	451	457	502	524	540
West Coast	935	1031	1005	1050	1363
Otago/Southland	2267	2331	2281	2809	4632
Total	**25818**	**28873**	**28873**	**32441**	**40403**

Source: Ministry of Forestry (1996)

Table 2.6: Estimated New Zealand Consumption of Roundwood ($000m^3$ RWE) 1981-1995

Year	Mean NZ Population (000s)	Total Removals	Imports	Exports	Total Consump.
1981	3147	10245	272	4904	**5613**
1982	3161	9753	348	4178	**5923**
1983	3190	9358	351	3880	**5829**
1984	3231	9266	422	4318	**5370**
1985	3259	9626	568	3911	**6283**
1986	3273	10195	656	3881	**6970**
1987	3282	9613	766	3670	**6709**
1988	3310	9688	566	4638	**5616**
1989	3318	10619	556	5630	**5545**
1990	3337	11744	626	6000	**6369**
1991	3373	13693	642	7876	**6459**
1992	3416	14136	700	9097	**5739**
1993	3452	14938	1070	9872	**6136**
1994	3491	15131	861	9119	**6873**
1995	3539	16437	996	10222	**7211**

Source: Ministry of Forestry (1996)

(1986) concluded that even under the most optimistic assumptions, demand for sawn timber cannot be expected to increase by more than 25 percent. While consumption of particle and fibreboard is expected to increase, the consumption of plywood is expected to remain remained largely static. Observing the figures confirms that for the past fifteen years per capita consumption of roundwood has remained largely unchanged.

The overall pattern evident from the demand and supply trends is clear. Given that supply is forecast to increase rapidly, while demand is expected to remain static, a large increase in the volume of roundwood available for export markets, directly or indirectly, can be expected.

Export Aspects

Relative Importance of Forestry Exports

Exports of forest products from New Zealand have in fact grown rapidly in recent years. Table 2.7 shows the comparative weighting of exports of forest products and New Zealand's other major export commodities in total merchandise exports for the years 1985 to 1995 inclusive.

Table 2.7: Value of Principle New Zealand Exports 1985-1995 ($NZ million)

Product	1985		1990		1995	
	Value	%	Value	%	Value	%
Meat Products	2288.8	20.78	2512.3	17.30	2753.8	13.63
Dairy Products	1434.8	13.03	2066.9	14.23	2747.5	13.60
Forestry Products	771.2	7.00	1361.2	9.37	2575.7	12.75
Wool	1475.4	13.40	1315.9	9.06	1252.9	6.20
Fruit & Vegetables	446.0	4.05	934.4	6.43	1195.2	5.92
Fisheries	366.2	3.33	697.0	4.80	1105.5	5.47
Fuels	-	0.00	515.1	3.55	919.3	4.55
Raw Hides & Skins	473.6	4.30	658.2	4.53	663.6	3.29
Textiles	102.9	0.93	174.7	1.20	291.8	1.44
Plastic Materials	100.9	0.92	126.3	0.87	261.8	1.30
Iron Ore & Concen.	42.0	0.38	21.1	0.15	22.4	0.11
Other	3510.1	31.88	4141.5	28.51	6410.3	31.73
Total	**11011.9**	**100.00**	**14524.6**	**100.00**	**20199.8**	**100.00**

Source: Statistics New Zealand (1996)

As can be seen, the forestry sector has become New Zealand's third largest single export sector. It can also be seen that exports from the forestry sector are rapidly approaching the level of dairy produce, New Zealand's traditional area of strength. Other traditional export products, meat and wool, have also declined in relative importance. The forestry sector has become of primary importance to the New Zealand export sector.

Value and Volume Trends

Turning next to the export trends within the sector itself. Table 2.8 shows total exports of forest products by volume and value, respectively, (in nominal dollars) from 1981 to 1995. Figures for annual percentage growth over the year before have also been provided.

Table 2.8: New Zealand Exports of Forestry Products by Volume ($000m^3$ RWE) and Value ($NZ000) with Annual Growth Rates 1981-1995

Year	Volume	Change (%)	Value	Change (%)
1981	4904	-4.5	532815	19.5
1982	4178	-14.8	550799	3.4
1983	3880	-7.1	503271	-8.6
1984	4318	11.3	651029	29.4
1985	3911	-9.4	796050	22.3
1986	3881	-0.1	767969	-3.5
1987	3670	-5.4	786296	2.4
1988	4638	26.4	985373	25.3
1989	5630	21.4	1237144	25.5
1990	6000	6.6	1385871	12.0
1991	7876	31.3	1577260	13.8
1992	9097	15.5	1824719	15.7
1993	9872	8.5	2323808	27.4
1994	9119	-7.6	2469055	6.3
1995	10222	12.1	2614629	5.9

Source: Ministry of Forestry (1996)

Looking first at the volume figures, the increase in the volume of exports of forestry products is clear, nearly tripling over the past two decades.

The rate of increase in the value of exports also stands out spectacularly. The sector has maintained an average annual growth in value of exports of around fourteen percent, with extraordinary growth displayed in certain years. In total, a twenty-five fold increase over the past two decades. It comes as no surprise that the sector is predicted to become New Zealand's number one export earner.

Export Destinations

Table 2.9 summarises the major markets for New Zealand exports by value over the period 1985 to 1995, and has been organised according to the major economic entities within the Asia Pacific region. All countries not specifically mentioned have been included in the "others" category. Two major patterns are evident from the data. The first is New Zealand's comparatively narrow export base. Export dependence on two major markets, Australia and Japan, is evident. Together these two countries account for approximately two thirds of the purchases of New Zealand's total forest products exports by value, with Japan recently becoming the largest single market (taking 29.9 percent of New Zealand exports by value in 1995). However, there is considerable difference between the two markets. Australia is a market for the more processed goods, while Japan (the main growth market) imports largely logs and woodchips (an issue developed further in the following chapters). Moreover, the Australian market has accounted for a declining proportion of exports over recent periods, and is expected to grow only slowly in absolute terms from now on, as their own forest resources come on line.

The second pattern evident is that the vast majority of exports go to the East Asian area, approximately 96 percent if we include Australia. In addition to Japan and Australia, other East Asian economies, notably Korea and Taiwan, also account for a significant proportion of purchases as single states (14.2 and 5.6 percent respectively). The East Asian Newly Industrialising Economies, account for over one fifth of purchases as a group, while the ASEAN-4 economies as a whole take approximately eight percent of exports by value (and have been identified by the Forest Industries Council as future growth markets). The value of exports to different countries within both of these groups does vary quite widely.

Exports to the North American region are small, not surprising given that both Canada and the United States are major wood producers. Also notable by their absence are exports to the European countries, a consequence of high levels of transportation costs at all stages of the industry.

Table 2.9: Value of Exports of Forestry Products From New Zealand by Destination 1985 - 1995 ($NZ000)

Region	1985		1990		1995	
	Value	%	Value	%	Value	%
Australia	**343014**	**45.02**	**550066**	**39.69**	**762431**	**29.16**
Japan	**173355**	**22.75**	**335201**	**24.19**	**781904**	**29.90**
EANIEs[a]	**61346**	**8.05**	**246120**	**17.76**	**615620**	**23.55**
Korea	20815	2.73	125212	9.03	371466	14.21
Taiwan	6046	0.79	65696	4.74	145417	5.56
Singapore	6646	0.87	22780	1.64	29378	1.12
Hong Kong	27839	3.65	32432	2.34	69359	2.65
ASEAN-4[b]	**68049**	**8.93**	**121101**	**8.74**	**204077**	**7.81**
Indonesia	19665	2.58	57894	4.18	108581	4.15
Malaysia	17745	2.33	20663	1.49	34098	1.30
Philippines	14095	1.85	17090	1.23	25128	0.96
Thailand	16545	2.17	25454	1.84	36270	1.39
China	**23185**	**3.04**	**21637**	**1.56**	**26870**	**1.03**
NAFTA[c]	**8401**	**1.10**	**24507**	**1.77**	**133468**	**5.10**
USA	7855	1.03	22981	1.66	131583	5.03
Canada	543	0.07	1526	0.11	1885	0.07
Mexico	3	0.00	-	0.00	-	0.00
Others[d]	**84603**	**11.10**	**87239**	**6.29**	**90259**	**3.45**
Total	**761953**	**100.00**	**1385871**	**100.00**	**2614629**	**100.00**

[a] East Asian Newly Industrialising Economies

[b] Association of South-East Asian Nations

[c] North American Free Trade Area

[d] Others are all other countries to which New Zealand has exported forestry products during the year.

Source: Ministry of Forestry (1996) and New Zealand Statistics INFOS EXI Series

Exports By Product

Perhaps the most interesting questions are raised when we examine exports by product. Debate is now being raised in New Zealand on the extent to which the New Zealand forestry sector is exporting raw materials rather than adding value to the product within New Zealand. Table 2.10 shows the composition of exports of forest products from New Zealand by value for the major product categories for the years 1981 through 1995. Clearly apparent from the table is the fact that the New Zealand forestry sector has a tendency to export less processed goods, where the classifications represent less to more processed as we move from left to right.

As was the case with the data on contribution to GDP covered earlier, the dominance of the less processed goods stands out quite clearly in the export data, with logs and poles and sawn timber, being the main exports. Particularly over recent years (since 1987) there has been a major increase in the value of log and pole exports, with the other products increasing at much slower rates. This would suggest that the current prosperity the industry is enjoying has been brought about largely by increasing sales of the raw product, rather than by adding value to the product in New Zealand.

Table 2.10: Composition of New Zealand Exports of Forestry Percentage of Value 1981-1995

Year	Timber Industry		Forest Products Industry				
	Logs & Poles	Sawn Timber	Wood Pulp	Paper Product	Panels[a]	Other[b]	Total
1981	9.3	16.4	28.4	28.7	5.4	11.9	100
1982	5.3	14.5	28.3	33.2	5.6	13.3	100
1983	5.6	14.4	32.1	26.8	5.5	15.6	100
1984	6.1	13.7	29.2	29.4	4.7	17.0	100
1985	4.7	15.9	25.5	29.7	4.0	20.2	100
1986	5.3	14.5	28.6	24.4	4.8	22.4	100
1987	5.3	11.9	32.2	22.5	7.9	20.2	100
1988	8.2	11.5	35.9	20.3	9.9	14.2	100
1989	10.5	14.4	32.2	21.2	10.8	10.8	100
1990	14.9	13.6	27.8	21.8	10.8	11.1	100
1991	20.9	14.2	24.7	19.3	11.3	9.5	100
1992	20.3	16.3	20.8	20.2	12.0	10.6	100
1993	30.1	18.5	15.2	15.6	11.7	8.9	100
1994	31.7	20.7	13.2	13.2	12.0	9.1	100
1995	23.7	19.7	18.4	12.7	15.7	9.7	100

a Panel Products is the total of fibreboard, plywood, veneer and particleboard.

b Other includes all forestry products not separately listed.

Source: Ministry of Forestry (1996)

Considering the export figures purely in terms of value may be somewhat deceptive, since the increase in log export values could have come about as a consequence of improving log prices per unit. However, the trend stands out even more clearly when we examine exports by product and volume. Table 2.11 does this. Looking at these figures it can be seen that the majority (67 percent) of the increase in total export volume since 1987 has been in low value-added logs and poles. This trend has coincided with the increase in foreign ownership of the resource.[6] Whatever the causes of the trend, it is clear that the transition to the higher harvest potential of New Zealand's forest resource has not been accompanied by the development of processing capacity. The trend is a major cause for concern in some quarters, and will be discussed further in the following chapter.

Table 2.11: Composition of New Zealand Exports of Forestry Percentage of Volume 1981-1995

Year	Timber Industry		Forest Products Industry			
	Logs & Poles	Sawn Timber	Wood Pulp	Paper Product	Panels*	Total
1981	16.4	26.7	36.9	22.9	4.4	100
1982	11.3	26.1	38.1	25.7	4.7	100
1983	11.3	24.9	41.6	23.9	4.3	100
1984	12.5	22.6	40.2	28.1	3.6	100
1985	9.2	28.1	37.2	25.7	3.1	100
1986	10.2	23.0	43.2	24.6	3.9	100
1987	11.6	21.2	43.5	21.8	7.1	100
1988	17.9	19.7	44.2	18.8	8.9	100
1989	27.4	20.1	35.5	18.6	8.3	100
1990	36.2	18.6	26.3	18.6	7.9	100
1991	41.8	17.3	24.8	14.3	7.5	100
1992	41.7	18.5	23.1	14.8	7.3	100
1993	48.0	19.6	20.0	13.8	7.8	100
1994	47.3	19.4	20.8	13.1	7.9	100
1995	47.0	20.4	20.4	11.4	9.9	100

* Panel Products is the total of fibreboard, plywood, veneer and particleboard.

Source: Ministry of Forestry (1996)

International Aspects

We now briefly discuss New Zealand's place in the world forestry products market as a producer and exporter of forestry products. Table 2.12 gives world production levels of industrial roundwood for 1993. Total world production in 1993 was approximately 1458 million m^3, of which 15 million m^3, was from New Zealand forests. In world terms New Zealand is a small producer, accounting for just over one percent of total production. Even if production levels nearly triple as predicted, and environmental concerns and over-exploitation of natural forests result in a decline in production in the rest of the world, New Zealand is unlikely to account for more than 4 percent of world production over the next 30 years.

However, forestry products, being relatively costly to transport, are generally traded only within a region. When considering the significance of New Zealand as a source of forestry products, it is misleading to consider the production levels of the entire world. It is the production pattern of the Asia-Pacific region that is of most relevance. Total production in this region in 1993 was approximately 1013 million m^3. Moreover, since hardwood and softwood markets can be regarded as separate, it is the softwood markets which are of most interest. According to FAO figures, approximately 65 percent of roundwood production in the region is softwood. Using this as a basis, New Zealand currently accounts for approximately 2.5 percent of production for this market, and may account for around 7 or 8 percent over the next 30 years. This level, while not huge, is not insignificant.

While New Zealand is not currently significant as a producer of roundwood, it is already a major exporter. In 1993 New Zealand was the fifth largest exporter of logs in the world, accounting for over 6 percent of total world exports. Given the magnitude of expected export supply changes, this figure could rise to as high as 18 percent by 2025, making New Zealand the second largest supplier of logs after the United States. In any case, New Zealand is already the world's largest exporter of roundwood originating from plantation timber (FAO 1993, p.155), and will almost certainly be among the three biggest log exporting nations next century. However, the same dominance cannot be observed in the processed forest products markets. New Zealand is only the eleventh largest exporter of sawn timber, and is positively minute when compared with countries like Canada. It fails to register at all in the paper and paperboard markets. New Zealand is predominantly a raw materials supplier.

In terms of the global political perspective, New Zealand is characterised as a moderate country. According to Brown (1997, p.42), its primary contribution in the international arena is often to act as an arbiter or conciliator, seeking ways to progress deadlocks between the more protagonistic countries. It also provides development assistance to the forestry sectors of the island states of the South Pacific, and plays an important role in international environmental initiatives, where its key aim is to ensure that plantation forestry as practiced in New Zealand, continues to be internationally accepted as a means of achieving environmental objectives (Brown 1997, p.42).[7]

Table 2.12: World Production of Industrial Roundwood 1993 (000m^3)

	Value	%	Value	%
Canada	173133	11.87		
United States	402500	27.60		
Other North America	11131	0.77	**586764**	**40.24**
Brazil	77808	5.34		
Chile	20534	1.41		
Other South America	21527	1.48	**119869**	**8.22**
China	100608	6.90		
India	24667	1.69		
Indonesia	39054	2.68		
Japan	25570	1.75		
Malaysia	44957	3.08		
Other Asia	34742	2.38	**269598**	**18.49**
Australia	17633	1.21		
New Zealand	15898	1.09		
Other Oceania	3371	0.23	**36902**	**2.53**
Finland	35483	2.43		
France	27673	1.90		
Germany	32361	2.22		
Poland	15962	1.09		
Sweden	49830	3.42		
Other Europe	82649	5.64	**253158**	**17.36**
Former USSR			**132355**	**9.08**
Africa			**59523**	**4.08**
World			**1458169**	**100.00**

Source: FAO (1995)

Summary and Conclusions

The forestry sector has become one of primary importance to the New Zealand economy. Its contribution to employment and GDP is large and growing, and it is a major export earner. However, it has been shown that the major increases in production volumes expected over the coming decades cannot be absorbed by the domestic market. The future of the forestry sector lies in exports. In this area two major points stand out. The first is the reliance of New Zealand on a few key markets (Japan and Australia). The second is the tendency to export unprocessed goods rather than add value to them prior to export. From an international perspective it can be seen that New Zealand is likely in the future to become a substantial exporter of forest products in global terms, despite the fact that it is an insignificant producer in global terms, and likely to remain so.

Notes

[1] Roche (1990) provides an excellent history of the forestry industry in New Zealand and its important role in the economy and society from the 1820s until the late 1980s.

[2] This Tribunal is charged with making recommendations to government over land and other grievances stemming from disregard to the principles of the 1840 Treaty of Waitangi, the signing of which heralded widespread European colonisation of New Zealand.

[3] For an analysis of the earlier years of the sales process see Swier (1993), and for a look into the background to the sales see Birchfield (1993).

[4] This figure, in that it does not include foreign shareholdings in the other large owners, is likely to underestimate the true extent of foreign ownership.

[5] Roundwood equivalent is a theoretical measurement unit giving the total amount of roundwood necessary for the production of one unit of a stated forestry product with existing technology as if only roundwood was used as a raw material.

[6] Brown (1997) also notes the correspondence between the log export increase and the privatisation of the forests, and puts it down to the private sector realities of obtaining returns on investment and maintaining liquidity (p.16).

[7] Debate is now being raised overseas as to whether the New Zealand model of 'timber farming' is environmentally sound. New Zealand's position is that since they prevent the use of natural forests, plantations of exotic species are a means of conservation. The position of some other countries (mainly in Europe) is that the introduction of exotic species is an environmentally unsound practice, and that only wood sourced from managed natural forests is environmentally sound.

3 Rationales for Processing Incentives

Introduction

In the previous chapter, the New Zealand forestry sector has been shown to be one which is not only of considerable importance to the New Zealand economy at present, but also one with significant future economic potential. However, it is not a sector which is without significant areas of debate, in particular in the trade policy area. In this chapter we deal more thoroughly with the separate but interrelated issues that have arisen in New Zealand and elsewhere with respect to the forestry sector and trade policy. We cover the issue of further processing, and the debate that has surrounded the desirability or otherwise of utilising export restrictions on unprocessed wood products to increase the level of processing occurring within New Zealand prior to export. We discuss what has become one of the focal points of more recent debate, the increase in foreign ownership of the resource. We deal with possible economic rationales for increasing processing, including what is often alleged to be the proximate cause of the low level of processing within New Zealand, the escalating tariff structures of the major importing countries of the region, and how these effectively shift processing from New Zealand to other competing economies. The possible impact of retaliatory actions is also considered, as are other potential arguments for domestic processing. We also briefly consider the legal environment for export restrictions and other processing incentives.

The Processing and Export Restriction Debate

As has been shown in the preceding chapter, one of the clear trends in the forestry sector of New Zealand is a strong and increasing dependence on exports of raw materials (logs). A question that is often asked is whether or not intervention to restrict exports of logs would be appropriate, and under what circumstances. To begin the discussion of this question, it is perhaps useful to examine the experience of the most well-known user of log export restrictions in the Asia-Pacific region, Indonesia.

The Indonesian Example

In 1978, Indonesia exported 72 percent of its forestry products output as logs. In 1980 the government banned log exports, thereby forcing the establishment of domestic plywood mills for processing. By 1986 over 100 mills were established and Indonesia's wood export pattern had changed drastically. It increased plywood exports from 70 thousand m^3 in 1978 to 4.6 million m^3 in 8 years. Japan, which was importing only 90 thousand m^3 of plywood in 1978, was importing 762 thousand m^3 by 1986, 94 percent of which was from Indonesia. By 1988 the relevant figures had rocketed to 1.9 million m^3 and 97 percent respectively. Wije-wardana (1989) likens the Indonesian case to what the "OPEC countries did for oil in 1973" (p.28).

The aggressive Indonesian tactic had a severe impact on the plywood industries of Japan and Korea; Japan's industry, with a capacity of 7.5 million m^3, sought government support; when it failed it had to scale down its production. Korea, the world's leading producer and exporter of hardwood plywood in the mid 1970s, had its export volume decline from 427 million m^3 in 1977 to 35 million m^3 in 1985. Industry capacity fell during the same period from 700 million to 370 million cubic metres.

In 1988 the Indonesian government went a step further in promoting added value exporting; large taxes were imposed on sawn timber exports, which were scaled down depending on the degree of further processing of the export product.

The Indonesian case is not, of course, without its critics. The policy has been criticised on both environmental and financial grounds. In place of an export ban, some economists believed an export tax on logs would oblige forestry concessionaires to move into downstream industries while also generating a new revenue stream for the government. Since log exports are banned and plywood exports are untaxed, the 'gains' from Indonesia using its market power have accrued to a relatively small number of private firms. The policy has also been severely criticised by some who have shown the loss in export earnings to be significant. Lindsay (1989) estimates the total losses at between 1.3 and 2.3 billion US dollars.

Despite these criticisms, policies to promote the processing of forest products prior to export are increasing in frequency. In the United States there has been an export ban on logs from state forests for many years, and according to the FAO (1993, p.156) almost all of the major Asian producing countries (including Malaysia, the Philippines) have introduced legislation restricting exports of unprocessed wood products. New Zealand's laissez-faire approach currently makes it the odd one out, although the free-market has not always ruled in New Zealand.

Trade Policy Intervention in New Zealand

New Zealand in fact has a long history of regulation and intervention in trade, and the forestry sector is no exception. New Zealand has until quite recently used a variety of trade and commercial policy measures to encourage various outcomes in the forestry sector. Here we briefly summarise measures which have been undertaken, and the rationales that were used to justify them at the time.

In the early period of New Zealand's colonisation, free trade was generally held to be the policy most beneficial to New Zealand's circumstances. Fleming (1990) notes how "New Zealanders saw the virtues of free trade as a powerful engine of growth transmitting a vigorous expansion for primary products from the industrialised centers of Europe and Great Britain to Australasia" (p.225). He notes how this entailed the view not that tariffs should not be used at all, but that "they shall be levied exclusively for revenue, not at all for protection" (p.225). The forestry industry was in general no exception to this rule.

The situation changed somewhat when in 1892 the Victorian State Government levied a tariff on rough sawn and dressed timber, while allowing baulk to remain being admitted free of duty. The move was generally seen in New Zealand as an attempt by the Australians to expand employment in its own sawmilling industry, while buffering it against foreign competition. An export duty was levied on exports of baulk from New Zealand in an attempt to counteract the Victorian tariff. The move was criticised on the grounds of its retaliatory nature, and in any case seemed to have little effect. New Zealand was at that stage not a particularly large supplier to the Australian market, which was dominated by imports from North America, and thus could not have expected much in the way of bargaining power.

The next major intervention in forestry trade occurred in 1901 for quite different reasons. The Timber Export Act of 1901 levied an export duty on round, squared and half logs and filtches. The move was promoted by the government and grew out of the concerns of the farming sector that the New Zealand dairy industry would suffer because insufficient kahikatea (a timber that was particularly suitable for butter boxes because of its non-tainting properties) remained for the export butter trade. Farming was at the time regarded as the principle industry that should be promoted in New Zealand, and the sawmilling industry was subjected to this political regulation in the 'national' interest.

In 1903 the export duty on all logs was raised. The reasoning for the increase was in stark contrast to the earlier reasoning given for implementing the duty in the first place. Mills, the Commissioner of Trade and Customs now argued that "the main thing for this colony is to conserve as much as possible [of] the timber industry for the benefit of our working men within our own boundaries" (Roche, 1990, p.158). Rather than keeping timber at home for the benefit of other important industries, or in an attempt to force the removal of harmful tariffs in foreign markets, the Timber Export Duty Act of 1903 is the first attempt at promoting the domestic processing of the forestry resource for the purpose of maintaining employment.[1]

During the period 1906-1908, mounting concern was expressed about the impact of North American Oregon pine imports. In 1908 a petition was presented to parliament urging an increase in the import duty on Oregon pine. A concurrent claim was that Oregon pine was being dumped in New Zealand at prices far below local rates. The free trade views of the early settlers to New Zealand began to be replaced with a more protectionist stance as competition from foreign sources became more intense. Government reaction to the situation was not confined to tariff policy, however. As early as 1896 the Railways Department began to be regularly approached by domestic sawmillers for reductions in railway freight rates to both encourage the timber export trade (by decreasing the costs of transportation to the nearest port), and for discounted rates for domestic timber to allow it to compete more effectively with Oregon pine (Roche, 1990, p.152).

In 1918 domestic price controls were introduced, in conjunction with a system of export quotas. The export quotas lasted until 1928 (domestic price controls were dropped in 1924). Export permits were required for the disposal of timber on overseas markets. The policy was aimed at "conserving New Zealand timbers for New Zealand use at prices based on the cost of production without regard to parity of price for exports" (Roche, 1990, p.163). This idea of maintaining a cheap domestic supply was in contrast to the earlier measures imposed on the industry, aiming (ostensibly) as it did to protect domestic consumers from movements in international prices rather than to protect producers from international competition.

Price controls were brought in again in 1936. Once more, the objective was to ensure a stable and cheap domestic supply of timber. The measures lasted significantly longer than the earlier experiment, eventually being dropped in 1965.

From the 1940s, state promotion of the pulp and paper industry began. The period was characterised by extraordinary levels of intervention (taking the form of direct subsidisation, building of infrastructure, and even direct involvement in the planning, design and implementation of the plans by the selected enterprises) at all levels in the industry, in the belief that only the state was capable of managing such an enterprise properly, and that premature private sector entry into the industry would be doomed to failure and would then discourage the future healthy development of the industry. The pulp and paper industry was at various times seen as strictly a domestic supply venture, and as a major potential export earner during the course of its institution. The primary objective of the government in promoting the industry seems to have been as part of a policy of import substitution, a belief in the promotion of further processing as a means of industrialisation. State ownership of large tracts of forests gave the government a tool to direct resources to processing.

During the period from 1962 until the early 1980s, emphasis shifted to the encouragement of exports, as it became clear that the plantations established earlier in the century would in fact be surplus to domestic requirements, and fears of a timber famine vanished. Tax breaks encouraged the planting boom of the late 1970s, and export incentives grew from the export market development scheme, introduced in the 1962 budget. These measures were not exclusively for the forest products industries, but did not exclude them. The measures were in part a reaction to the 1961 Australian implementation of an export market development taxation incentive scheme, and in part to poor economic performance through the 1950s and the threat of Britain joining the EEC leading to a perceived need to rapidly diversify and expand exports. The moves taken include measures to reduce income taxes either by reducing assessable income or providing tax credits, such as the IETI (Increased Exports Taxation Incentive) and EMDTI (Export Market Development Taxation Incentive), and measures to encourage the growth of productive capacity, such as the EMIA (Export Manufacturing Investment Allowance). Most of the schemes (in particular in later years) targeted value-added.

From 1985 came the deregulation of the industry and the removal of export incentives. In line with the general policy thrust of the Fourth Labour Government, the state removed itself from a position of control over the industry, selling its forestry assets. The move represented a coming full circle, and a return to the free trade ideals held in New Zealand during the last century. Export restrictions are not used for plantation forests in any form, although the export of indigenous woodchips and unprocessed logs is banned under the Forest Amendment Act unless it is felled in an area under a sustainable management plan. This bill was, however, brought in for environmental reasons rather than concerns over the level of processing. In any case the level of production from native resources is minimal. Subsidies are also not in use, with the exception of a $15 million conservation forestry scheme on the East Coast, which has been in place since 1994, again for environmental purposes. However, despite the current dominance of laissez-faire in New Zealand policy and business circles, the debate over processing and export restrictions has never quite died away altogether.

The 1993 Resurgence

The debate over export restrictions became a major topic of debate again in 1993 when it was reported that imports of logs from Chile were being considered because of a shortage of logs reaching domestic sawmillers. The incident highlighted a growing frustration among sawmillers over the increasingly high domestic log prices that resulted from an open export market, although there were also allegations that forest owners were refusing to supply local sawmillers even at international prices. The controversy grew further with the involvement of the Council of Trade Unions, which complained that sawmiller's jobs were under threat as log supplies dried up because of the higher prices paid in lucrative Asian export markets. The timing of the resurgence of the export restriction debate corresponded to a large 'spike' in log export prices (which has since tapered off somewhat).

Next to enter the fray was a newly established group of approximately 100 sawmilling and timber processing companies called the New Zealand Owned Sawmillers Group. This group was quick to pursue the export restriction line, with a report alleging that the industry would collapse without limits on the export trade. The report called for a quota system to guarantee an increasing volume of wood for domestic processing, requiring log exports to be restricted to 36 percent of all log production, and each log exporting company to make at least 18 percent of its total log sales available to New Zealand-owned sawmills (by this stage there were a number of foreign-owned sawmills). They also published the findings of a Heylen survey of 1000 people, which found that 90 percent of New Zealanders wanted log exports limited in favour of domestic processing.

While the arguments of the New Zealand Owned Sawmillers Group seemed to touch a chord of public sentiment, the reaction of the forest owners and government was negative, and the possibility of export restrictions was dismissed. Price setting was to remain a commercial matter, and was to be resolved only between the interested parties. It was made quite clear that the government did not intend to intervene.

The resurgence of calls for export restrictions was not altogether surprising in an environment of increasing log export prices squeezing the profit margins of New Zealand's sawmillers. Similarly, the polarisation of views on the subject between forest owners (who perceived that they stood to lose from such a move), and the sawmillers (who clearly stood to gain) was not surprising. What was missing from the debate was a comprehensive attempt to model the effects of export restrictions and their likely order of magnitude. While few would advocate a return to the interventionist excesses of the past, there are legitimate questions to be answered with respect to the benefits of policy induced increased domestic processing. Is it possible to increase economic welfare at the same time as increasing processing? Is restricting exports an appropriate policy intervention to increase the level of processing, or are other policy interventions likely to be less damaging in welfare terms? These are questions that remain to be addressed.

Potential Economic Rationales for Further Processing

Foreign Ownership of the Forest Resource

The sale of New Zealand's state-owned forests as detailed in the previous chapter was extremely controversial. Part of the controversy lay in the belief of some commentators that the sale price was too low, and that the government had virtually given away a highly valuable asset. Others saw the issue in more philosophical terms, arguing that the sale represented a loss of New Zealand sovereignty. While the issue of sovereignty is a difficult one to analyse from an economic perspective, there are several important and legitimate questions involved here.

While few economists would disagree that foreign investment can under certain circumstances be beneficial for the recipient country, the issue is not clear cut. Foreign investment is, from a balance of payments perspective, not all that different from overseas borrowing. Both are merely inflows of capital necessary to balance current outflows. In either case, the inflow of capital generates a claim on the income stream that flows from the underlying assets, which then accrues to foreign interests. Even if the company chooses to reinvest the income resulting from their investments back into the recipient rather than repatriate the money to their home country, it is not immediately obvious that the recipient benefits, since in this case more claims are generated.

If we consider foreign investment as merely a different form of overseas borrowing, it becomes clear why we can regard it as neither necessarily good nor bad. The question can only be answered by considering each case individually. Where it is likely to be beneficial is where there are spillover effects to other areas of the economy, where the investment leads to the introduction of new technology (physical or management) that would not otherwise have been available, or where the investment opens previously closed markets. Where it is unlikely to be directly beneficial is where the investment is merely a transfer of ownership of an already existing resource (unless the foreign owners are, for some reason, more efficient). Of course, where trade flows are not free, the movement of factors may lead to a more efficient allocation of resources that improves global welfare, but not necessarily the welfare of the recipient. Foreign ownership does not lead to any expansion in productive capacity or any lifting of gross capital formation in and of itself. While it may be argued that, due to the particular character of the forestry industry, ownership of a section forest resource is a prerequisite for investment in downstream processing, true economic investment will occur only when the latter investment takes place.

While a complete analysis of the impacts of foreign direct investment is beyond the scope of this study, the existence of substantial levels of foreign ownership of New Zealand's forest resource does have a significant and little recognised impact on the issue of export restrictions. It is well known that trade policy interventions are a means (albeit an inefficient one) to effect transfers of income. An export tax, by shifting income from forest owners to processors, may effect a transfer of income from foreign to domestic sources. How does this affect optimal export restricting strategies?

Terms of Trade Effects

For the trade economist, perhaps the most natural argument for restricting exports of logs, quite aside from the fact that it may increase processing, is the fact that doing so may increase the price of logs. It is well-known that by imposing an export tax a large country may force its firms to exercise the monopoly power which they (as small, competitive agents) do not realise they possess, to improve the country's terms of trade, and thus national welfare. This is simply the 'optimal' tax argument.[2]

In fact, there is a considerable body of literature concerned with the nature of terms of trade effects and the forestry industry. Much of the literature is concerned with the possible impact of the so-called 'feedback effect', which refers to the extent of increased foreign demand for processed domestic wood products resulting from foreclosure of the raw material source, i.e., the idea that increasing log prices may in fact flow into increasing processed good prices, creating an extra incentive to restrict exports. We deal in more detail with this concept in the following two chapters.

Tariff Barriers Faced by New Zealand

Perhaps the most commonly expressed argument for encouraging domestic processing of forestry products is the fact that other countries continue to maintain high levels of protection against processed products. Despite the commitments for reductions in tariff barriers that have occurred following the Uruguay Round, the forest products sector remains among the most protected in the Asia-Pacific region. Figure 3.1 depicts the current tariff levels. The main importers of the Asia-Pacific region are Japan, Korea and China, with Japan by far the largest. However, the tariff structures of all of these countries are designed so as to encourage the importation of raw materials and less processed products, and to discourage the import of more processed goods, i.e., they are of the so-called "escalating" tariff structure.

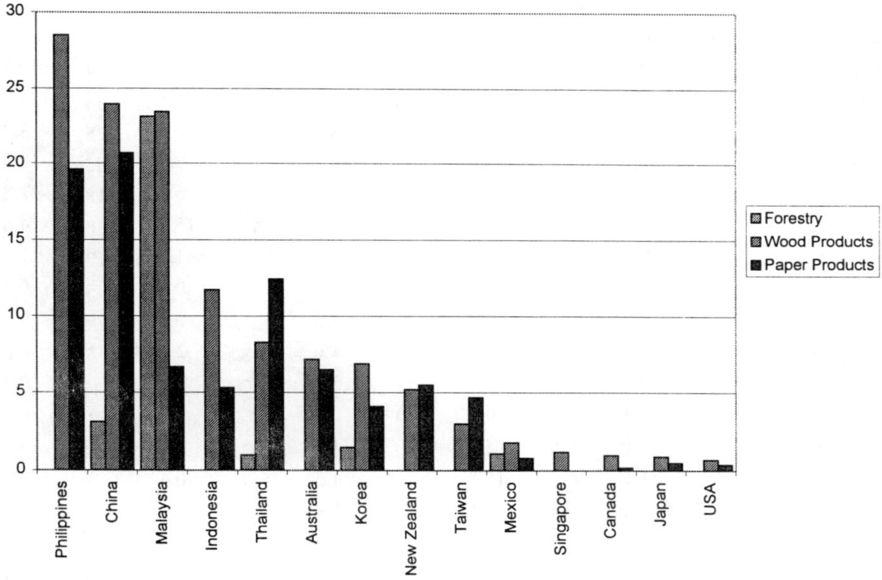

Figure 3.1: Weighted Average Tariffs (1995)
Source: McDougall et al. (1998)

While we note that it is not certain whether the Japanese demand for raw materials rather than processed products stems entirely from government decision or market dictates, this argument is referred to frequently by domestic sawmillers as the cause of the country's inability to export more value-added products.

An examination of the New Zealand export data broken down by both product and destination (Table 3.1) confirms the existence of a geographically based difference in the level of processed goods exports, with Japan, Korea and The Peoples Republic of China in particular standing out as being large importers of unprocessed logs and poles (and in the case of Japan also woodchips), while importing only very small quantities of the more processed goods.

Table 3.1: New Zealand Exports of Forestry Products by Destination and Product by Volume 1995 (Provisional)

Destination	Logs & Poles (million m^3)		Sawn Timber (million m^3)		Wood Pulp (000 tonnes)		Paper Products (000 tonnes)		Panel Products (million m^3)	
	Vol.	%	Vol.	%	Vol.	%	Vol.	%	Vol.	%
Australia	6	0.1	457	41.8	98	14.8	208	59.5	153	23.3
Japan	1916	39.9	285	26.1	238	35.8	1	0.3	324	49.1
EANIEs[1]	2716	56.5	123	11.3	126	19.0	78	22.3	158	24.0
Korea	2609	54.3	43	4.0	54	8.1	-	0.1	25	3.8
Taiwan	106	2.2	66	6.0	71	10.7	-	0.0	85	12.9
Singapore	1	0.0	1	0.1	-	0.1	23	6.4	18	2.8
Hong Kong	-	0.0	12	1.1	1	0.1	55	15.7	30	4.6
ASEAN-4	27	0.6	21	1.9	156	23.5	40	11.5	12	1.9
Indonesia	-	0.0	-	0.0	109	16.3	3	0.9	5	0.8
Malaysia	5	0.1	1	0.0	15	2.3	21	6.1	-	0.1
Philippines	17	0.4	6	0.6	11	1.6	8	2.4	5	0.8
Thailand	5	0.1	14	1.3	21	3.2	7	2.2	2	0.3
China	87	1.8	-	0.0	19	2.9	-	0.0	1	0.1
NAFTA	55	1.1	173	15.8	10	1.5	2	0.6	-	0.1
USA	55	1.1	172	15.8	10	1.5	2	0.6	-	0.1
Canada	-	0.0	1	0.1	-	0.0	-	0.0	-	0.0
Mexico	-	0.0	-	0.0	-	0.0	-	0.0	-	0.0
Others	1	0.0	33	3.0	17	2.5	20	5.8	9	1.4
Total	4806	100.0	1091	100.0	665	100.0	349	100.0	659	100.0

Source: Ministry of Forestry (1996)

This data would seem to confirm at least in part the sawmiller's view of the problem, and few could deny that an escalating tariff structure in place effectively discourages New Zealand from producing higher value-added products for export to Japan and the other countries. It can be reasonably assumed that the ratio of raw to processed forestry products would be lower if the escalating tariff structure was not in place. However, we know from elementary trade theory that the pattern of trade is based on the resource endowments of the trading countries. As Wiseman and Sedjo (1981) point out, at least in the past, the tendency of the Japanese to rely heavily on log imports rather than on imported timber "reflected Japan's comparative advantage in labour-intensive, wood processing activities *augmented* by

Japanese trade policies directed towards the protection of the processing industries" (p. 424, emphasis added). For this reason, it is not clear to what extent the situation would change were the so-called problematic countries to reduce their tariffs. The tariff of Japan on radiata sawn-timber for example is 4.8 percent. Unfortunate, but not exactly huge. How much effect does it really have? This is a question that needs to be addressed from an empirical perspective. We do this in Chapter 8 of this study. It is also interesting to note that forestry products have been included in the APEC Early Voluntary Sector Liberalisation (EVSL) initiative, which aims to fast-track liberalisation measures within APEC on a sector-by-sector basis. Since most of New Zealand's forestry export markets are members of APEC, the implication is that New Zealand may soon be facing substantially lower trade barriers in forest products. The possible impact of the EVSL initiative is also examined in Chapter 8 of this study.

Another question raised by the existence of escalating tariff structures is whether or not there is anything New Zealand can do about it. It is a considerable jump from the fact that foreign escalating tariffs discourage New Zealand processing, to the conclusion that New Zealand should utilise measures to encourage it. The approach of New Zealand so far has been to utilise the GATT and other negotiations to try to get these countries to reduce their barriers. However, if New Zealand is in a situation where it provides a significant proportion of the primary input to the protected manufacturing sectors overseas, and if it is in a position to force up the price of the factor input, this may under certain circumstances be a welfare improving strategy for New Zealand, as well as a means of reclaiming the processing that has in effect been 'stolen'.

Returns to Scale and Dynamic Arguments

This study concentrates on the static impact effects of trade policy in the context of models that feature perfect competition, and constant returns to scale. There are a number of potential arguments for processing that cannot be adequately considered in this framework. Perhaps the most well-known dynamic argument for protection is the 'infant industry' argument. This is in fact one of the oldest arguments for protection, being first proposed by the then Secretary to the US Treasury, Alexander Hamilton in 1791, and popularised by Friedrich List. Essentially the argument goes along these lines. A firm cannot compete if it is small; it has to be large before it can become competitive. Therefore it has to be protected or assisted for some time, and be permitted to grow. Once growth is completed, assistance can be withdrawn. The argument is thus based on potential or long-run comparative advantages (the country cannot realise its true comparative advantage with other countries already entrenched in the sector).

There are a number of reasons for doubting the whether the infant industry argument is valid. Essentially, the argument rests on the assumption that the assistance leads to the dynamic expansion of the production possibility frontier (whereas static analysis considers only movements around the frontier). We need to carefully consider the basis for this expansion. Generally it is argued that expansion comes about due to economies of scale of some sort. Here there are two counter-arguments. The first is theoretical. The fact that there are economies of scale does not provide an argument for assistance, since if expansion of an industry were to genuinely offer high future returns, why do private investors not build the industry themselves? It is sometimes argued that investors only consider current returns, but this is not consistent with market behaviour, and New Zealand is not a developing country without functioning capital markets. Implicit in the infant industry argument is therefore the dubious notion that governments know better.[3]

The second argument is empirical. We reject the hypothesis of increasing returns to scale in the forestry processing industry in econometric analysis undertaken in Chapter 6. Furthermore, while it may be argued that external economies (i.e., social benefits that are not captured by the private investors – making them unwilling to invest in the industry) justify assistance, we consider this unlikely in the case of the New Zealand forestry sector. Such arguments rely on the existence of technological spillover effects from investment. Turning logs into timber is not rocket science, the technology is very standard. It is therefore difficult to imagine that substantial spillover effects exist.

Even if we believe that the infant industry argument is relevant for the New Zealand forestry sector, a study of the current type is still important. One of the criteria for evaluating whether or not infant industry policies are justified in any particular case is the Bastable's test. This criterion is that the assisted industry must be able to pay back the national losses incurred during the period of assistance. In other words, the discounted sum of the future benefits under free trade (when assistance is removed) must be at least as great as the discounted sum of the deadweight losses incurred. In order to apply this criterion it is necessary to know the magnitude of the deadweight losses, which the quantitative section of this study addresses in Chapters 7 and 8.

More recent economic research into trade policy has emphasised the importance of scale economies for a different reason. Scale economies tend to imply imperfect competition, and there are a number of models incorporating imperfectly competitive markets that shed light on various trade policy issues. These models have become known as 'strategic' trade policy models, and have been popularised by such authors as Helpman and Krugman (1989), and Krugman (1987). It may be argued that there can be some sort of strategic gain from restricting log exports, or other processing incentives. Once again, we emphasise that the empirical evidence presented in this study does not support the hypothesis of increasing returns to scale at the aggregate level, and we therefore believe that strategic type arguments are not very relevant in the current context. Moreover, there are a number of substantial criticisms of the strategic trade policy literature, including the extreme sensitivity of the models to the underlying assumptions, their partial equilibrium nature, the difficulty in identifying strategic industries, and the lack of empirical evidence to support the arguments. Perhaps more importantly, New Zealand is a small player in international markets, and forest products are commodities for which there are many substitutes and competing suppliers. For the country that takes international prices as given the domestic market structure is not relevant, since the solution for all possible market structures will be replicated by the perfectly competitive case so long as only price mechanisms are utilised. We therefore, while recognising the existence of dynamic and strategic arguments for protection, nonetheless consider the traditional neo-classical approach to be the most appropriate.

Transport Cost Arguments

It is often argued that as the level of processing increases, the cost of transportation decreases. This argument has been used to justify using export restrictions to increase domestic processing, as in Lin (1993). The evidence is somewhat less than clear on the matter. Considerable evidence (Roemer, 1979; Wall, 1980; Yeats, 1981; and Takeuchi, 1983) suggests that transport costs in fact escalate with the degree of processing – a phenomenon reflecting pricing to what the load will bear.

In fact, we would argue that not only is the evidence sketchy, but that whether or not transport costs rise or fall with the level of processing may not be directly relevant anyway. If the transport costs are constant for raw materials and processed goods then whether they escalate or not is irrelevant. To say that transport costs differ for each good is then the same as saying the price differs for each good (which is obviously true in general), and to argue that this means we should process more or less is clearly spurious. If there are scale economies involved in the transport costs, then we return to the same arguments presented above for the case of economies of scale in production, and once again there is no general justification for assistance to increase the level of processing. Hence, we consider the transport cost argument to be of limited applicability.

Other Possible Effects of Export Restrictions

In addition to the welfare and processing implications which form the focus of this study, there are a number of other reasons which have been proposed to justify the use of export restrictions. While most of the reasons fall beyond the scope of the current study, and some are of fairly limited applicability to the current situation in New Zealand, we briefly mention other the possible justifications here:

- Assurance of continuity of domestic supply, and prevention of shortages caused by abnormal foreign demand. This rationale has been used mainly by developed economies, notably Canada in the case of forestry. This justification has also been used in the case of New Zealand, as discussed in previously.
- Stabilisation of export earnings and protection against fluctuations in commodity supply, a policy more typical of developing economies.
- Response to aggregate balance of payments surplus problems – used by the oil exporting countries to reduce inflationary pressures generated by accumulated export earning surpluses (Weinblatt 1985, p.16).
- Conservation of non-renewable resources. This is a policy more appropriate to a natural resource than to the plantation forestry of New Zealand.

The details of these and other justifications for export restrictions are discussed at further length in Weinblatt (1985).

The Legal Perspective

Export Restrictions

While this study concentrates largely on the economic implications of export restrictions and other processing incentives, it is also necessary to briefly comment on the legal status of such measures under international trade agreements.[4] According to McGovern (1995), the terms of the General Agreement on Tariffs and Trade (GATT) reflect the view among most WTO members that the most significant threat of interference with international trade comes in the form of import restrictions rather than export restrictions (p.5.24-1). Nevertheless, export restrictions are covered by the GATT, with the most substantial requirements being consistency with the MFN and non-discrimination obligations of Article I:1 and Article XIII. These regulations clearly state that export restrictions must be applied to all countries in a non-discriminatory manner, if used at all. Export restrictions are also covered by the regulations concerning quantitative restrictions in Article XI, which are prohibited in general under the GATT. Hence, an export quota would be illegal, whereas an export tax would not. Technically, the provisions of Article II allow for the binding in export taxes, but the possibility has not been significantly exploited (McGovern 1995, p.5.24-1). Exceptions to the prohibition of quantitative restrictions are covered in Article XI:2, where they are applied to relieve or prevent critical shortages of products essential to the exporting country (temporarily) or where necessary for the application of standards or regulations for the classification of commodities in international trade. Other general exceptions are covered in Article XX, in the following circumstances:

- Article XX(g) is an exception for measures relating to the conservation of an exhaustible natural resource, if they are made effective in conjunction with restrictions on domestic production or consumption.
- Article XX(i) permits restrictions on exports of domestic materials necessary to ensure the availability of essential quantities of such materials to a domestic processing industry. While this is probably the section of the GATT which the NZOSG had in mind when it made its request for a log export quota system, the restrictions can only be implemented where the price of the affected goods is held below world prices as part of a governmental stabilisation programme. Moreover, there is a specific provision such that any such restrictions may not operate to increase the exports of the processing industry.
- Article XX(j) permits action to be taken when it is essential to the acquisition or distribution of products in local short supply.

Since Article XX(i) explicitly excludes the case where export restrictions operate to increase exports of a processing industry, it seems unlikely that either a log export quota system, or an export ban, would be legal under the GATT. Article XX(g) is the most likely candidate for the other export restricting countries (e.g., Indonesia) to use in their defence. However, the fact that the New Zealand resource is made of exotic plantations may present some difficulties for New Zealand in this respect. If export restrictions were to be implemented in a manner consistent with the GATT rules, they should be in the form of duties, and applied in a non-discriminatory manner.

Subsidies

Subsidies are covered in Article XVI of the GATT, with countervailing measures being covered under Article VI. However, these provisions have been largely surpassed by the more comprehensive WTO Agreement on Subsidies and Countervailing Measures. The agreement classifies a subsidy as being prohibited if it is specific to an enterprise or industry (Article 2) and is contingent on export performance, with specific examples being contained in Annex I. These provisions effectively make the use of export subsidies in any form difficult. Other specific subsidies, such as processing subsidies, while not prohibited, may nevertheless be classified as actionable under Article 5 if they cause injury to the domestic industry of another member, or cause serious prejudice to the interests of another member. Serious prejudice is defined as existing under Article 6 where there is a total ad valorem rate of subsidisation of greater than 5 percent, or subsidies that cover operating losses being sustained by an industry. Serious prejudice may also be deemed to exist if the subsidy has the effect of displacing exports of a like product from a third country market, or if it results in a significant increase in market share. However, these last two categories cannot arise when the complaining member has restrictions on imports of a like product. Therefore, for example, Japan could not claim serious prejudice on these grounds if New Zealand subsidised production of wood products, since it maintains tariffs on wood products.

In summary, export subsidies are prohibited under the WTO rules, and therefore cannot be utilised to promote domestic processing. A direct processing subsidy is allowed, but may be actionable if it can be shown to cause serious injury to overseas competitors. Given the size of New Zealand in global forestry terms, this seems unlikely to be the case.

Summary and Conclusions

This chapter has identified three important issues with respect to an export restricting trade policy and the New Zealand forestry sector. These are the effect on processing and welfare, the scope for income redistribution from foreign interests and how this effects optimal trade policies, and finally the impact of foreign tariff structures on New Zealand processing and the potential of retaliatory measures. Some of these issues have received significant media comment, and have become long-standing and important matters of public debate. Despite this, there has been little effort made to either analyse the implications of trade policy intervention in a comprehensive manner, or to quantify the likely effects.

Notes

[1] Maintaining domestic employment, particularly during seasonal downturns, was for many years seen as one of the roles of the Forestry Service, and was one of the factors (in addition to fears of an imminent timber famine) behind the push for plantation of exotic species on a large scale.

[2] It should be noted that the optimal export tax, like the argument with respect to foreign ownership above, relies on a transfer of wealth, rather than wealth creation, and it imposes overall losses on the world economy.

[3] Wall (1980) argues that the economies of scale argument with respect to processing is a red herring along similar lines (p.313).

[4] A more detailed discussion of the GATT regulations in respect of export restrictions is contained in Rom (1985) and Rom (1985b).

4 Forestry Models

Introduction

The objectives of this chapter are twofold. The first is to review the applied overseas literature relevant to the issues identified in the previous chapter, the main focus of being on modelling literature. There is a variety of related literature emanating from the large forest products producing countries of the region, in particular Canada, the United States, and Indonesia. Given that the focus of this study is on trade policy issues, we do not go into any great detail on the specific economic problems unique to forestry (estimation of supply, optimal rotation periods, etc.). These issues are better covered elsewhere.[1]

The second objective of this chapter is to discuss economic modelling in and on forestry in New Zealand. Overall, this chapter serves as a review of the relevant domestic and overseas applied research on trade policy and processing issues, and to place this study in context with the existing literature.

A Review of Forest Sector Trade Models

Forestry models can be broadly classified into three main categories: supply models, sectoral models, and multi-sectoral models. The first category is concerned exclusively with the estimation of supply of the raw material, and the optimal management of forest plantations. At its most simple level, this category is represented by the crop-yield table. At higher levels it also concerns such issues as optimal rotation periods, based on classical theory developed by Faustmann, and optimal control theory. Of more direct relevance are the sectoral and multi-sectoral models, as it is at this level that trade issues come into play. Sectoral models can be distinguished in a number of ways, including the level of disaggregation of inputs and outputs, the treatment of dynamics and regions, economies of scale, and behavioural criteria. The three main types of sectoral models are dynamic simulation, linear programming, and spatial equilibrium. Of the three, it is the spatial equilibrium approach that is used most often in studies similar to this one, but we briefly discuss the nature of the other types in the following section.

Multisector models explicitly attempt to capture the interactions between the forestry sector and the rest of the economy. These are discussed further later in the chapter.

Modelling Approaches

A dynamic simulation model is a model built on the assumption that there is a set of dynamically interacting decision bodies (subject to constraints of technology, resources, etc.) that determine a development trajectory. This gives rise to a system of difference equations that form the simulation model. The approach is useful for considering the long-term development of the forest sector. An example of the use of such models is Lönnstedt (1986), designed to analyse the impact of cost competitiveness and wood availability on the structural change of the forest sector over a 20-30 year period in Sweden. A problem with the approach is the fact that the equation system has to be restricted in terms of both functional forms and parameter values in order to obtain a well-behaved solution trajectory. This problem has been addressed by the use of dynamic linear programming models, which also require constraints on the initial and terminal states, and are solved by maximising an objective function. This solves many of the equilibrium and stability problems of the simulation approaches.

Although simulation and programming approaches can be applied to regional analysis of the forestry sector, and thus to trade, a more popular method is the spatial equilibrium approach, which is better suited to application to market (as opposed to command) economies as a means of determining an efficient allocation of production to satisfy demand in various regions.

Conceptually, the spatial equilibrium approach is quite simple. Figure 4.1 illustrates. In this case we have defined our "world" as two countries, *A* and *B*. Both countries produce two products, logs (*R*) and lumber (*L*). We express lumber in terms of its roundwood equivalent. *QA* is therefore quantity in country *A*, in log units, similarly *QB*. In the diagram country *A* has a lower autarky price for lumber than country *B*, leading to opportunities for gains from trade. Equilibrium is where the excess supply of lumber from country *A* (*XSL*) is equal to the excess demand for lumber in region *B* (*XDL*), *PL*. The physical requirements for lumber production in each country thus determined, the excess supply and demand for logs in each country follows, and the equilibrium price and quantity is determined in the log market as well.

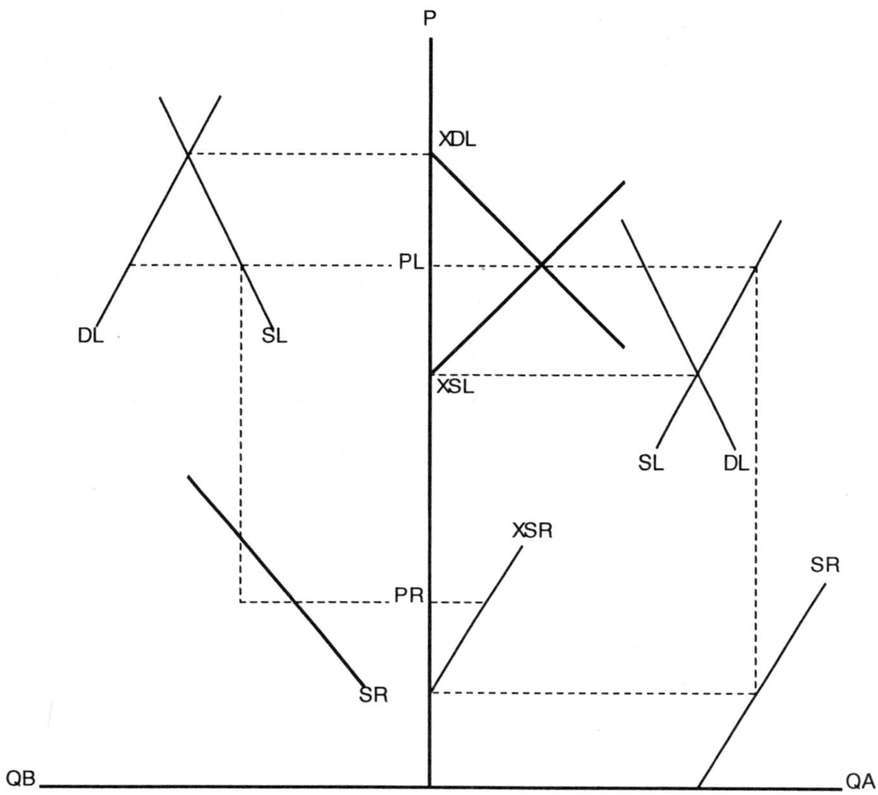

Figure 4.1: Spatial Equilibrium

Market Interactions - The Feedback Effect

A number of spatial equilibrium models have been inspired by threatened United States tariffs on imports of Canadian softwood. In the late 1980s the measures under consideration were to impose voluntary or mandatory quotas, tariffs, or to redefine the term 'subsidy' in existing countervailing duty law to increase the probability of favourable administrative rulings. Boyd and Krutilla (1987) utilise a spatial equilibrium model of the North American lumber market to estimate the production and welfare impacts of various U.S. trade restrictions on Canadian lumber. Their model incorporates 34 supply regions and 39 demand regions, in an attempt to fully capture the regional character of the North American lumber market. They use their model to simulate a variety of potential tariffs and quotas.

Broadly, their findings are that the losses incurred by Canadian producers in the event of a U.S. tariff would be substantial, but that voluntary export restraints by Canada could lead to gains as high as 40 percent of pre-existing profits.

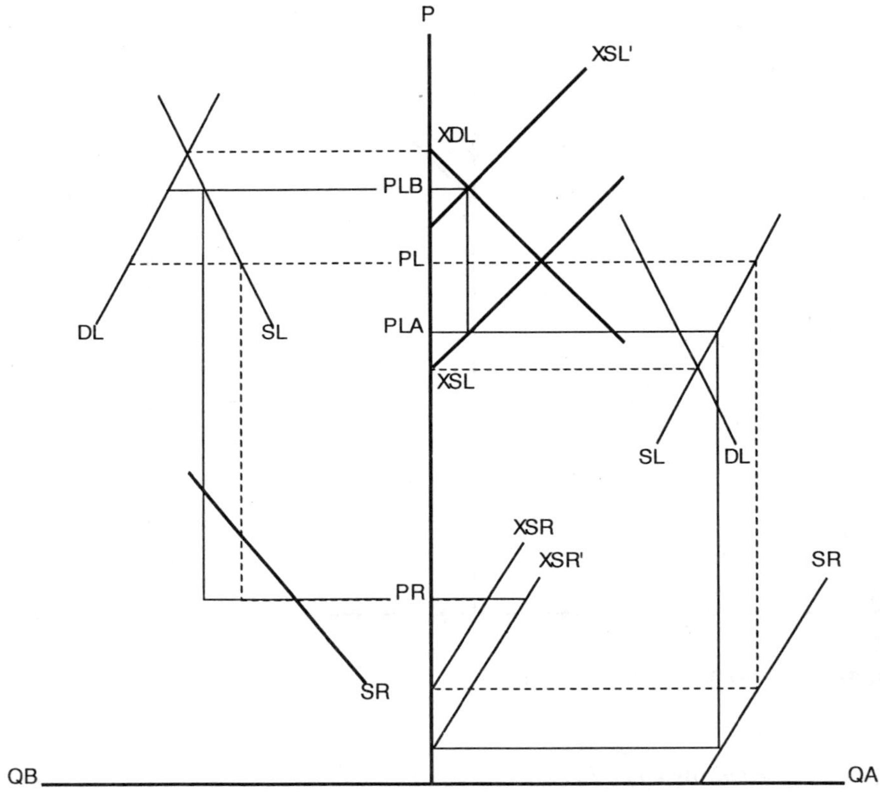

Figure 4.2: Impact of a Tariff on Lumber

Boyd and Krutilla (1987) apply their spatial equilibrium analysis to the lumber market only, and so do not consider the impact trade restrictions applying to lumber may have on log exports (or on exports of other processed goods) and the modifications to welfare effects which may result. To see why this may be important, consider the impact of a tariff in the simple diagrammatical framework given above (refer to Figure 4.2). A tariff on lumber shifts the excess supply curve for lumber upward to, say, *XSL'*, increasing the price in country *B* (say, the United States) to *PLB*, and decreasing the price in country *A* (say, Canada) to *PLA*. Production in Canada declines, and production in the United States increases. This is the effect that the Boyd-Krutilla (1987) model is capturing. However, the policy will also result in an expansion of the excess supply of logs from Canada to *XSR'*, and of the excess demand for logs in the United States. Exports of logs will increase (although the impact on price and welfare is ambiguous). Ideally, when considering the effect of trade policy on forest products, these sorts of interactions should be taken into account. This gives us a strong reason to use methodology involving the simultaneous treatment of both markets.

Moreover, this is not where the interactions end. We can also consider the impact of the so-called 'feedback effect', which we make a slight deviation here to discuss. The possibility of feedback was first debated by Weiner (1973) and later Haynes (1976) and (1977), and refers to the extent of increased foreign demand for processed domestic wood products resulting from foreclosure of the raw material source. Wiseman and Sedjo (1981), identify the limits of the feedback effect in a paper that applies the Marshallian derived demand model in the context of international trade, to derive estimates of the market equilibrium, net welfare, and welfare incidence effects of a hypothetical ban of softwood log exports from the Pacific Coast region of the United States.

Broadly speaking, the concept of feedback can be thought of in the following way. We have in our diagrammatical treatment assumed that the lumber supply curves are fixed independent of the supply of logs. However, given that logs are the major input into lumber, changes in the price of logs should impact on the supply of lumber. Consider Figure 4.3. An export tax on logs lowers the price of logs in the exporting country and raises the price in the importing country (*XSR* shifts up). This is effectively a subsidy to lumber production in the exporting country, and a tax on lumber production in the importing country. We would expect the tax to result in a contraction of supply in the importing country. This results in an upward shift in the excess demand for lumber. This is the feedback effect. Note that the shift in supply functions feeds back into the demand for logs. A new equilibrium will eventually be reached where the price paid for logs is higher in the importing than in the exporting country. The restriction of exports of logs unambiguously causes an expansion of lumber exports if there is feedback (note however, that the tax also causes the excess supply curve in the exporting country to expand, so the effect on lumber price is ambiguous; in the diagram it is unchanged).[2]

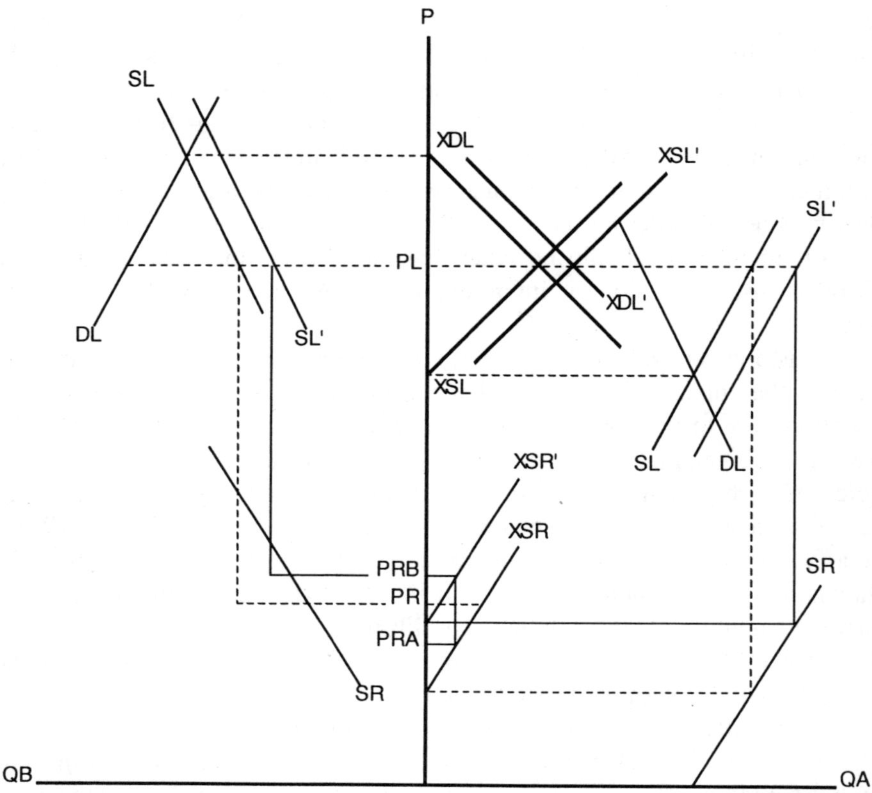

Figure 4.3: Impact of a Log Export Tax with Feedback

Wiseman and Sedjo (1981) use a single country model (facing a downward sloping demand for exports). They are unable to consider the extent of feedback in terms of the supply schedule of the foreign country, but are forced to illustrate the effects of a log embargo for two limiting cases, which they define as "no feedback" (i.e., where the foreclosure results in no shift in the excess demand curve) and "maximum feedback" (i.e., the case where the increase in demand for lumber is equivalent to the amount of logs formerly exported).

In the no feedback case their results are such that if foreign demand for lumber is independent of export log sales, a log export embargo will reduce both the price of logs and lumber for the home country. Compared to the initial situation with no export control, welfare losses exceed welfare gains. The prohibition results in losses to log producers and gains to domestic processors and buyers of lumber. Some of the total gain to buyers accrues outside of the country (as a result of the fall in the price of lumber).

In the case of maximum feedback, the home log price still falls, but by less, and the price of lumber rises. This results in losses to log producers, gains to producers of lumber from both lower input prices and higher prices of their output, and losses to consumers. The result may or may not be a net welfare gain. The results can thus be summarised as follows. For any feasible theoretical feedback level, the imposition of an export embargo will result in the falling of the domestic price of logs and their production, and the owners of resources used in log production suffer losses. The reverse is true for the processing industry, which experiences gains in both limiting cases. Regional buyers of lumber win or lose, depending on the price impact of the feedback effect on the lumber market (which may rise or fall). The region as a whole may realise a net welfare gain or loss, depending on the size of the feedback effect.

A spatial equilibrium model with multiple products in multiple countries avoids the problem of being only able to consider limiting cases. There have been a number of such models built. A spatial equilibrium model has been built to analyse international trade in pulp and paper (Buongiorno, 1986), and the Timber Assessment Market Model (TAMM) was developed to facilitate long-range planning and policy analysis in the US forest products sector (described in Adams and Haynes, 1986). The TAMM approach was later expanded to support analysis of how developments in offshore forest products markets would influence US forest resources, producers, and consumers. This model is known as the World Assessment Market Model (WAMM), and is described in Brooks (1987).

The most well-known of the large-scale models built along spatial equilibrium lines with a global focus is the Global Trade Model (GTM). The GTM started life at the International Institute for Applied Systems Analysis, but was taken over by the Center for International Trade in Forest Products in the late 1980s. A complete description of the model structure can be found in Kallio et al. (1987) and Cardellichio et al. (1989). The current version projects production, consumption, prices and trade for ten forest products in 43 timber supply regions, maximising the value of the sum of consumer and producer surplus. The model is dynamic, taking the equilibrium results from a base year and using them to find equilibrium solutions for subsequent years, taking into account changes in demand, production and trade levels. These changes are implemented through sub-models for timber supply, production capacity, and consumption.

The CGTM has been applied to a fairly wide variety of issues important to the global forest sector. Studies of most direct relevance to the present study include Flora and McGinnis (1989), Lippke (1994), Lippke and Perez-Garcia (1992), Perez-Garcia (1991) and (1993), and Perez-Garcia et al. (1994). Of these papers, Flora and McGinnis (1989), Lippke and Perez-Garcia (1992), and Perez-Garcia et al. (1994) deal specifically with the issue of export restrictions (in the case of Flora and McGinnis the removal of export bans), while the other papers deal with issues relating to curtailment of supplies due to environmental impacts or other factors. The results of all the papers are quite similar, and so here we concentrate on Perez-Garcia et al. (1994).

This paper analyses the prospect of an export ban imposed on timber produced on private land in the Pacific Northwest. The model suggests that a log export ban would eliminate 12.5 million cubic metres in log exports by 1995. The export ban raises log prices in the rest of the world, while lowering prices in the export ban region. The primary log-consuming markets of Japan, Korea and Taiwan-Hong Kong suffer log price increases.

The log export ban simulation results imply the substitution of logs from Chile and New Zealand for the restricted logs from PNW. The increase in log exports from these regions is not, however, enough to make up for the shortfall, and significant log price increases result. Given the greater difficulty faced by Asian processors in obtaining logs, production of lumber in Asia declines by more than production increases in PNW. Asian countries are forced to import more lumber, largely from Canada. The policy results in a decline in welfare for the United States, borne by domestic producers of logs and consumers of lumber, the loss is partially offset by gains to the lumber processors. Domestic consumers of lumber face a decline in welfare as a result of an export ban or tax due to increased lumber prices.

As an alternative, Perez-Garcia et al. (1994) considers the possible impact of a log export tax that maximises revenue. If the tax revenues are included, the United States gains overall from the policy. The distribution effects and the effects on the rest of the world are the same as with an export ban, although milder in effect (since an export tax only restricts part of the log exports, not all of them). Of course, this gain comes at the expense of the rest of the world (as the authors point out), and may be eliminated by retaliation.

Other spatial equilibrium studies have been applied to Indonesia. Sidabutar (1988), for example, builds a model including logs, lumber and plywood to compare the export ban imposed by Indonesia with the counterfactuals of non-intervention and a less restrictive log export policy.

He concludes that the combined export revenue from logs, lumber and plywood would have been larger under a less restrictive policy, and that the current policy is welfare reducing. Interestingly, he also concludes that plywood export revenues would be larger the less plymill capacity was expanded. This result is similar to that of Lindsay (1989), who shows that the log export restrictions in Indonesia resulted in an expansion of plywood exports, pushing the price of plywood down. This is a case where the feedback effect is not sufficient to offset the additional loss from what is effectively an export subsidy to plywood manufacturers in the presence of some level of unexploited monopoly power.

There are a number of weaknesses in the approach taken by the CGTM and other spatial equilibrium models that are worth briefly discussing here. The first is that spatial equilibrium models do not reproduce actual trade flows with much accuracy, although this is not of major concern if the model is to be used for policy analysis. Perhaps more importantly, Cardellichio and Adams (1987) argue that with the GTM the specification of economic behaviour in the model is too rigid for long-run simulations that involve significant changes in endogenous variables. The only economic adjustments that occur in the model endogenously are short-run adjustments in final product demand and delivered wood supply. This implies that the model will not give reasonable results in simulations involving major economic or policy shocks.

Finally, the use of fixed proportions in resource use in this class of models is somewhat restrictive. The use of fixed proportions may imply an overestimation of the extent of any feedback effect, as overseas processors are unable to substitute away from logs when prices increase, which in turn implies an understatement of potential price falls in the processed good market.

Production Technology and Vertical Integration

The assumption of fixed proportion technology common in spatial equilibrium models also means that changes in wood-labour-capital costs have no effect on the optimal choice of production technique. In fact, the possibility of input substitution may open up new rationales for increasing processing that spatial model are unable to capture. Lin (1993) utilises an idea first put forward by McKenzie (1951), and later developed by Vernon and Graham (1971) and Shughart (1990), to show the importance of production technology in providing a vertical integration rationale for increasing processing where there are interrelationships in production. In order to illustrate the consequences of production technology and vertical integration, we consider a highly simplified example, from Lin (1993).

Assume that there are two successive stages of production of a good (lumber - *L*). In the first stage an intermediate input (logs - *R*) is harvested, and in the second is used in combination with labour to produce the final good. We assume initially that the production technology is fixed proportions (i.e., only one input combination will be able to produce the desired level of output), and that there is no close substitute available for the intermediate input. Logs are supplied from only one source, while the world market for lumber is perfectly competitive (i.e., there are many lumber producing countries and many firms within each country, all of whom are price takers). The country that is a monopoly supplier of logs has already set prices so as to maximise profits. The question we want to answer is will the input monopolist step into the downstream processing industry in order to increase its profits?

Since lumber is produced and sold under perfectly competitive conditions, the marginal cost of lumber, MC_L, is equal to the sum of the price of logs, P_R, and the transformation cost MC_T (i.e., $MC_L=P_R+MC_T$). The marginal cost of harvesting logs is given by MC_R. Units are defined such that one unit of logs is required to produce one unit of lumber (i.e., $Q_L=Q_R$). For simplicity, we assume that both marginal transformation cost and the marginal cost of harvesting are constant.

Because the lumber industry is perfectly competitive, each lumber producing country and firm will produce the quantity of lumber such that the marginal cost of production is equal to the given price of lumber, thus $P_L=MC_L=P_R+MC_T=MR_L$, where *MR* is marginal revenue.

The monopoly supplier of logs selects the intermediate input quantity Q_R so that $MR_R=MC_R$, and sells this quantity at price P_R to maximise profits. The monopoly profits are therefore $(P_R\text{-}MC_R)Q_R$. Since $P_R=MC_L\text{-}MC_T$, and since $MC_L=P_L$, we can rewrite the monopoly profits as $(P_L\text{-}MC_T\text{-}MC_L)Q_R$.

If the input monopolist vertically integrates with the downstream processing industry, its profits can be written as $P_LQ_L\text{-}MC_TQ_L\text{-}MC_RQ_R$. Since $Q_L=Q_R$, the profits become $(P_L\text{-}MC_T\text{-}MC_L)Q_R$, which is exactly the same as profits obtained without integration. We can thus conclude that there is no incentive for the upstream monopolist to integrate with the downstream industry as long as the downstream industry remains competitive. Even if the monopolist were to acquire the entire downstream industry (an extremely unlikely scenario), what would happen is at best still charging P_L for the final good, and supplying Q_L. Therefore, the only consequence of integrating the entire downstream industry is that the monopoly supplier would collect its profits from final good sales rather than from intermediate input sales.

Now let us consider what would happen if the production technology allows variable proportions, i.e., where the input combination choice for a desired level of output depends on the relative prices of those inputs. In order to incorporate the impact of variable proportions technology on our analysis, we utilise the graphical demonstration of Vernon and Graham (1971). We assume that lumber is produced by two distinct inputs, logs and skilfulness of labour, S. The contribution of capital is embodied in the skilfulness of labour. A lumber producer can use either more skilful labour (or better equipment) or more logs to achieve an increase in production. Logs therefore have a substitute once increasing log prices become a problem for a lumber-producing, log-importing country. We maintain our assumption that logs are provided from a single monopoly source, while the lumber market is perfectly competitive. We also assume that the supply of labour is perfectly competitive.

Figure 4.4 describes the situation. If the supply of logs (R) is monopolised, while labour (S) is supplied competitively, then the price of labour P_S will be equal to its marginal cost MC_S , while the price of logs P_R will exceed it marginal cost MC_R. The corresponding isocost curve in this situation is given by *C0C0'*, the slope of which is $-P_R/MC_S$. *X0* and *X1* are isoquants of lumber production. If we assume that the production function is homothetic, then the slopes of the isoquants are invariant along any given ray from the origin, which then represent the factor intensity of production. To produce the quantity *X0* of lumber, producers will choose the input combination represented by the point *B*. If both inputs have been provided competitively, however, the lumber producers should face isocost curves like *C2C2'* or *C0C0"*, the slopes of which are $-MC_R/MC_S$. In other words, if both inputs are supplied competitively, it costs only *0C2* (in terms of units of labour inputs) to produce the same output *X0* (point *C*), or alternatively, the same cost (*0C0*) could produce the higher level of output *X1* (point *A*). The isocost curve *C1C1'* through point *B* with slope $-MC_R/MC_S$ shows us that if lumber producers faced the competitive prices, but used the wrong input combination *B*, then it costs more to produce *X0*. The vertical distance *C0-C1* represents the cost imposed by the upstream monopolist on the downstream processing industry (attributable to monopoly pricing of logs). It also represents the profits that are earned by an unintegrated upstream monopolist. The vertical distance *C1-C2* represents the cost to the downstream industry of the use of an inefficient input combination.

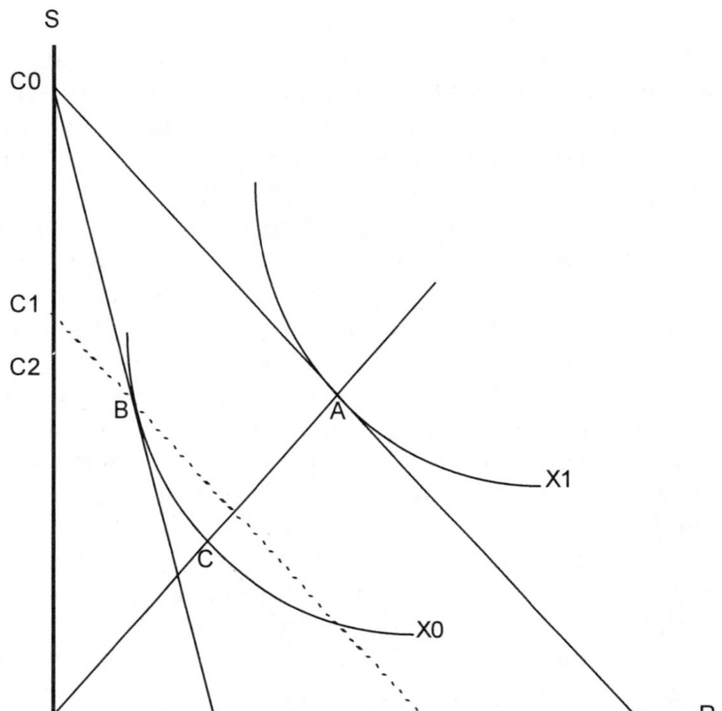

Figure 4.4: Variable Proportions and Processing Incentives

Without integration, the downstream processing industry uses factor proportions that under-utilise logs (as a result of substituting away for them due to the higher price). If the monopolist engages in processing, it can gain extra profits *C1-C2*, since under common ownership of stages of production, the monopolist would transfer the logs to the final good division at marginal cost. The downstream industry, which now faces competitive prices, would now choose the correct input proportion. The vertical merger allows the monopoly supplier to correct the inefficiency in factor proportions.

Of course, there is a crucial difference between a country with monopoly power over its exports and a monopoly firm. For the country as a whole exports are supplied by a large number of small firms, not by a single large one. This is why an export tax is necessary to force the firms to act in unison and take advantage of the monopoly power which they, as individuals, fail to recognise. The export tax distorts prices both at home and abroad (raising the price abroad, and lowering it at home). Since the owners of the raw materials industry are not necessarily also the owners of the processing industry, the processing industry does not face the competitive price, but the new distorted domestic price. The implication is that logs will be over-utilised in production in the country that imposes export restrictions. This in turn implies that a combination of taxes on exports of the raw material, taxes on domestic usage of the raw material, and taxes on exports of the final good (if monopoly power exists for that good as well) is necessary for a truly welfare optimal processing strategy. This position is unlikely to find favour among those promoting export restrictions for the raw material alone.

Overall, the implication of the above type of analysis for the problem at hand is that variable proportions technology in forest products processing can provide an extra incentive for a large supplier of raw materials to move into processing, and that this effect is ignored in models that assume fixed proportions technology. An important question is then, is the technology in wood processing fixed or variable proportions? In fact, utilising more capital intensive production techniques, or more skilled (and thus expensive) labour, may well increase the output of certain products for a given level of wood inputs, particularly at the more dissagregated levels. We present some econometric evidence in Chapter 6.

Multiple Sector Models

The problems involved with allowing for factor substitution and the effect this has on trade and welfare, in particular with respect to the feedback effect, point to the need for general rather than partial equilibrium modelling techniques. Moreover, in addition to problems in dealing adequately with intrasectoral linkages, there may be important intersectoral linkages as well. It may not be realistic to treat a wide range of economic variables as exogenous, since, at the national level, forest sector policy will have consequences for other industries.

There are a range of possible approaches to dealing with intersectoral linkages, the most common of which are input-output (IO) models, linear programming (LP) models, and computable general equilibrium (CGE) models.[3] IO models have been in common use since the 1950s. However, while IO models do capture sectoral linkages to a limited degree, they have some serious limitations. It can be argued that they are in fact more suited to command economies, since they fail to capture the market mechanism. The theoretical underpinnings of such models, in particular fixed relative output and factor prices and perfectly elastic factor supplies (i.e., the lack of primary factor constraints), also tends to reduce their usefulness in cases involving shocks to a sector accounting for a large share of economic activity. A further important shortcoming relates to the treatment of international trade in these models. Exports are exogenous, and imports are treated as non-competing. Hence trade does not depend on relative prices, and the crucial issue of the change in trade flows in response to price changes is not accommodated.

LP models, briefly discussed above, were developed in the 1960s to incorporate choice and optimisation into policy models. The results of LP models can be regarded as approximations to general equilibrium solutions (Dervis et al. 1982, p.64). However, despite the fact that LP models have been able to incorporate primary factor and trade constraints, they have not been successful in handling them in a realistic manner, often being overly sensitive to minor changes in relative prices (Bandara 1989, p.50).

CGE models are increasingly the approach of choice for economic policy analysis in market economies. These models use computer based solving techniques to perform general equilibrium calculations, and attempt to replicate the workings of the market economy by simultaneously solving for market prices and quantities. The technique has the significant advantage over other techniques that it maintains all income identities at all times. Since this is the methodology chosen in this study, a more detailed consideration of the CGE methodology is delayed until Chapter 6, so that it can be discussed alongside the development of the model utilised. Here we consider two previous models of the forestry sector that have been built along computable general equilibrium lines.

The first model is from Percy and Constantino (1987). The model is a small-scale CGE model applied to British Columbia and Canada. It is specified in proportional change form, assumes fixed supplies of factors of production, and makes use of the Armington specification of imperfect substitution in consumption between domestic production and imports. The model is of a single country, and considers the timber industry, the wood products industry (both broken into two regions - Interior and Coast), the paper products industry, the mineral sector, domestic manufacturing, and domestic services. It is used to consider the impact of an export tax on lumber imposed by the Canadian government in 1986 (of 15 percent) with the revenue directed back to British Columbia, and the possible impact of a hypothetical 15 percent duty imposed on British Columbia lumber exports by the United States. Both policies are shown to have similar effects. Log prices fall, as do the prices of wood and paper products. Output of logs and wood products fall, while output of paper products remains largely constant. The model also captures a wide range of complex intersectoral consequences of the policies. For example, the duty reduces output of logging and wood processing industries, releasing labour and putting downward pressure on nominal wages. The import-competing goods sector (domestic manufacturing) and the mineral sector both expand as a result. These results would be missed in a partial equilibrium analysis.

The CGE model used by Lin (1993) is applied to the Indonesian case. The model is single country, and consists of five sectors (logging, wood processing, non-wood agriculture, non-wood manufacturing, and non-tradeables). The model allows for substitutability of primary factors in wood processing by utilising a two-level constant elasticity of substitution (CES) production function. Capital and labour endowments are fixed, and capital is assumed to be sector-specific. The model also makes use of the Armington assumption. Other assumptions are quite standard and very much in the neo-classical tradition. Savings is always equal to investment, the government is assumed to redistribute all revenues back to households, and preferences are homothetic. Imports are assumed to be available at fixed prices (i.e., for imports Indonesia is treated as a small country), while downward sloping demand schedules are specified for exports (to allow for the impact of market power). In contrast to Percy and Constantino (1987), the model is implemented in non-linear form.

The model is applied to the estimation of optimal export taxes on logs and plywood simultaneously. Lin finds that an export tax of 45 percent on logs combined with an export tax of 21 percent on plywood is optimal. The extent of export restrictions on logs also of course affects the appropriate level of taxes on processing, Lin also finds that if log exports are banned altogether, a tax of 54 percent on plywood is optimal. Taxes that maximise tariff revenue are also calculated.

Finally, the optimal response to an increase in tariff faced in the processing industry is considered. Lin shows that, for a large producer of logs, the optimal response to a restriction imposed in the processed goods market is to tighten restrictions on exports of logs (provided they do not already exceed the optimal level), regardless of the level of monopoly power held in the processed market.

The works of Percy and Constantino (1987) and Lin (1993) represent significant improvements over the spatial equilibrium methodology used in most studies of the forest sector for cases where the forest sector forms a large part of the economy under analysis. The multi-sectoral specification allows for the representation of intrasectoral and intersectoral linkages in a way that is not possible with partial equilibrium techniques.

Both studies are of a single country, and while this is entirely appropriate if the country is small, in both of these models foreign demand is specified as a separate constant elasticity of demand function for each product (i.e., both authors make the simplifying partial equilibrium assumption that the foreign demands for net exports of each good are a function of own prices). Since the excess demand function of the processed good is not in any way linked to the world price of logs in either model, the possibility of feedback is ignored entirely. In both of these models the price of the processed good falls with an export tax on logs, in contrast to the results obtained in similar studies using the CGTM, discussed above. A fall in the lumber price is certainly a possibility, but the specification of these models means that it is a certainty. Unless combined with a tax on lumber, a log export tax can only lower or leave unchanged the price of the processed goods. In other words, by ignoring the feedback effect, both models may overstate the extent of processed good price falls resulting from restriction of log exports. Ideally, a CGE model should be built in a way that models in the feedback effect in an explicit fashion, if the country of interest is large enough to influence world prices.

Studies of Trade Liberalisation

While the above sections have largely focused on the issue of export restrictions, there are also a few studies which have concentrated on the impact on processing of foreign liberalisation of escalating tariffs. One of the first studies along these lines was Golub and Finger (1979), which analyses the effect of the removal of developed country tariffs on processed goods on processing and welfare in developing countries. Their simulations suggest that processing in the LDCs would increase by 8.3 percent, and decrease by 0.9 percent in the DCs, with the removal of all barriers to trade. This provides evidence to support the hypothesis that escalating tariffs have the effect of transferring processing from raw material exporting to raw material importing countries.

There have been relatively few studies where the focus of the trade liberalisation analysis is specifically on the forestry industry and processing. UNIDO (1983) estimated that the complete removal of tariffs in ten developed countries would have an expansionary effect on trade in wood products (increasing wood and wood products trade exports for developing countries by 3.3 percent). However, this study groups forestry and processed products into the same category. Another study by the FAO (1988) addresses the effect of the removal of selected Japanese tariffs on forest products, for Indonesia, Malaysia, and the Philippines. The study estimates gains from increased exports of processed wood product exports of between $3 million (for the Philippines) and $10 million (for Indonesia). Both of these studies used partial rather than general equilibrium methodology. No studies have been made of the impact of similar measures on the forestry industry of New Zealand.

Modelling Forestry in New Zealand

With respect to the first category of models, the supply planning models, New Zealand has a long and largely respectable history. The first modelling was done in the early 1920's. The first national supply planning model was prepared by Familton (1969), and was subsequently revised by Hosking (1972), Levack (1979), Elliot and Levack (1981), and the New Zealand Forestry Service in 1982. These models comprised regional supply forecasts, which were then aggregated to form a national statement.

One of the earliest attempts at broadening the scope of sectoral planning for forestry was the Forestry Sector Strategy Model (FSS - also known as the Ministry of Works and Development (MWD) Forestry Model). This model was developed jointly by the New Zealand Forestry Service and the Ministry of Works and Development to systematically explore the future of the forestry sector. A full description of the model and its structure is contained in Grant et al. (1978) and (1979).

The strategy model is a dynamic simulation model, and is essentially an instrument for carrying out "what-if" inquiries against a variety of assumptions. The model takes a starting snapshot of conditions as they existed in New Zealand in 1970/71, from which new levels are calculated for the next time period according to explicit theories about the behaviour of investors, and given values for the growth in national population, income, industrial output, productivity, etc.

At least one attempt has been made to link the FSS model to a larger, economy-wide model. Lowen and Philpott (1980) join the FSS with the Victoria Project on Economic Planning Model (PEP). The PEP is an economy-wide linear programming model (a full description of the PEP model can be found in Philpott and Elley, 1974 and 1974b). The linking of the models consists of using consistent estimates of economic parameters from the PEP model as parameters in the FSS model (with some modification to ensure material balance consistency in the forestry sector) to obtain predictions on production and export volumes. While the model approach can be argued to incorporate the influence of other sectors on the forestry sector, the interaction occurs in only one direction. This situation is unlikely to be appropriate for a sector as large as forestry in New Zealand, which is likely to have important general equilibrium influences on the rest of the economy.

In the 1980 forest industry study commissioned by the Development Finance Corporation it was recommended that much more attention be directed to the modelling of the marketing, manufacturing, transporting and harvesting elements of the sector, and far less to wood supply alone. This led to what Whyte (1989) refers to as the "scenario modelling approach", whereby models such as IFS, GROHA, MOVPRO and RSUM, predicting the effects of implementing regional scenarios for growing, harvesting, transporting and selling forest products, were summed to arrive at a national plan. While these refinements represent a significant improvement beyond the simple supply based models, some authors, such as Leslie (1986), made it clear that they were still did not constitute adequate forest sector planning. Certainly, the framework left little room for policy analysis. The same approach was taken in the Central North Island Planning Study (CNIPS) in 1983. This study was widely criticised for failing to present any alternatives to the single "plan" that was offered.

Summary and Conclusions

In conclusion, on modelling the New Zealand forestry sector two main points stand out. The first is that, as Hunter (1987) points out, New Zealand models have in the past been "excellent in the sense of being domestic raw material supply predictors, but are just passable in terms of domestic demand estimates" (p.40). The second is that the trade and policy component of forestry sector modelling has traditionally been a weak point. All New Zealand studies that currently exist are qualitative in nature.

There are a number of extensions possible. It is possible to analyse trade policy issues using partial equilibrium methodology provided the sector is too small to have any significant impact on real output, as a number of overseas studies reviewed in this chapter attest. These models may be multi-country or region (commonly referred to as spatial equilibrium models). Recognition of the impact from and to other sectors is, however, generally limited to the use of broad economic aggregates (output, population, housing starts, etc.) in the derivation of demand coefficients. Given the size of the New Zealand forestry sector, it can be argued that it is only at the multi-sector level, utilising general equilibrium methodology, that meaningful analysis of trade policy issues can take place in an internally consistent manner.

One approach might be to link an existing single sector model to multi-sector models representing the entire economy, and thereby introduce the influence of other sectors through factor movements and demand movements. This approach, however, raises the spectre of methodological and data inconsistencies, which are likely to be extremely time-consuming and difficult to eliminate. An alternative, if the objective is policy analysis, is to abstract away from the details of the forestry sector models, and deal with the broader policy variables impacting on the sector and the economy as a whole. This is the approach, using CGE methodology, that is taken here.

Notes

[1] For an overview of the theoretical framework in forestry models see Johansson and Löfgren (1985) and for examples of their application see Kallio et al. (1986).

[2] This concept of 'feedback' is clearly a matter best dealt with in a general rather than a partial equilibrium framework. This is done in the following chapter.

[3] See Robinson (1989) and Dervis, de Melo and Robinson (1982) for surveys of multi-sectoral models. A good example of the application of IO models to processing issues (although not in the forestry sector) is Bocoum and Labys (1993).

5 A General Equilibrium Approach to Processing and Trade

Introduction

In this chapter we develop a competitive general equilibrium model to deal in a more complete manner with the issues that were outlined in the preceding chapters. The main objective of this chapter is to set out a formal analytical framework for analysing the impact of processing incentives in a trade-focused, general equilibrium setting, which we will use as the basis for interpreting the results of the larger, more complex computable models that follow. In particular, the questions we wish to ask of the model are as follows. Under what circumstances can a country simultaneously exporting a raw material and a processed good benefit in terms of increased economic welfare from using export restrictions on raw materials to increase domestic processing? What are the impacts of such a policy on the incomes of the owners of different factors of production? What are the impacts on the production structure in the economy? And finally, how does the presence of foreign owned factors of production impact on these questions? For some of these questions we are unable to provide unambiguous answers, but there are a number of insights into the nature of the problem that can be gained from the process. The modelling thus also serves as a means of highlighting the issues for which an empirical approach using a numerical general equilibrium model is required.

The chapter can be broadly broken into four sections. In the following section approaches to the issue of export restrictions and processing in the theoretical literature are briefly reviewed. We then turn to the development and analysis of a small-scale, competitive general equilibrium trade model for the small country (price exogenous) case. Considerable space is devoted to the analysis of this case both for the clarity in the progression of results that it allows, and because of the obvious importance of the small country case for an economy the size of

New Zealand. We then proceed to consider the implications of modelling the issue of processing incentives in a true general equilibrium setting, making use of a two country model. Throughout we utilise the duality approach of Dixit and Norman (1980), Kohli (1991), and others, as this simplifies the mathematics of the problem considerably. Finally, we consider the impact of generalising the model to incorporate many goods and many factors, presenting the results we obtain in the preceding sections in the light of the generalised model of Woodland (1982), of which our models (and those others discussed in the following section) can be thought of as special cases. A summary and conclusions follow at the end of the chapter.

A Brief Review of the Literature

In the previous chapter we have reviewed some of the applied models that have attempted to deal with issues similar to the ones at hand. Here we briefly review the works that form the theoretical foundations of the approach taken in this study.

Traditional models of trade as embodied in the Heckscher-Ohlin-Samuelson (HOS) model and others of its ilk, are based on what is known as the "*classical paradigm*". That is, their fundamental assumption is that international markets are limited to final goods, while inputs are trapped within a nation's boundaries. This sort of approach has two major weaknesses when it comes to analysing real world patterns of trade. The first is that these models allow only for inter-industry trade, and ignore the large volume of intra-industry trade. The second major weakness, and the weakness which is of primary importance in this study, is that no allowance is made for simultaneous trade in both primary goods and the final products into which they can be processed.

There are a number of different ways in which we could model trade in a resource such as forest products. Lin (1993), as part of her study of Indonesia, approaches the problem by using a model with two final goods, one of which is a processed form of the raw material (plywood), and the other an aggregate importable. The raw material (logs) is treated as a fixed endowment, and can thus be interpreted as being equivalent to a tradeable factor of production (all other factors of production remain untraded). The model is in fact based closely on the works of Woodland (1983) and Svensson (1984), which deal with mobile capital. Treating logs as a factor with a fixed endowment makes the development of the model somewhat easier, and could be interpreted as a fixed annual allowable cut (as does exist in some countries). However, it means that the production choices in the raw material industry are independent of price, and to this extent the model is likely to overstate the effectiveness of policies designed to encourage domestic processing, since total output of the raw material cannot fall.

Perhaps the most natural direction from which to initially approach the issue of trade in forest products is the literature on trade in natural resources. Much of this work was pioneered by Vousden (1974), and Kemp and Long (1984) provide a review of the literature. They argue the need for a theory of international trade that accommodates both exhaustible natural resources and the traditional Ricardian primary factors, and of which the standard Heckscher-Ohlin theory appears as a special case. Much of the literature in this area is concerned with optimal rates of extraction over time of a non-renewable resource (although some authors do deal with self-renewing resources - such as Tawada 1981 and 1982). The models tend, by their very nature, to be much more focused on the dynamic aspects of the extraction problem than the static analysis of trade. The models, like Lin (1993), treat the raw material simply as an exogenous given, with the difference being that it is not a fixed annual endowment, but rather a fixed initial endowment that declines dynamically.

Although an application of the literature on trade in natural resources seems initially appealing, it is difficult to justify the use of such models in the analysis of New Zealand forestry. It is quite clear that the literature on this area is more suited to the analysis of minerals and fossil fuels, which are genuinely exhaustible, than to forests. While it could certainly be argued in the case of a country like Indonesia that the tropical hardwood forests can be realistically treated an exhaustible and exogenously given resource (albeit slowly self-renewing), this is not the case in New Zealand where the resource in question is in fact a privately owned and developed plantation forest. In the case of New Zealand it would seem much more appropriate to treat the supply of logs as what it is, a competitively produced commodity. (Lin, 1993, in fact does this in her CGE model, though not in the theoretical model on which she bases her work). It may be argued that it takes time for a tree to grow to maturity, and therefore the decision on whether or not to plant is independent of price at harvest. If this is the case it may be argued that there is therefore effectively a fixed endowment at time of harvest. However, this ignores the fact that there is considerable flexibility involved in the choice of when to harvest. Trees may be cut when still quite young for pulpwood, or left to grow further if prices fall. Also, planting is only one part of the production process. The silviculture and logging processes, both of which have a significant impact on the overall value (and volume) of the final product, should also be taken into account and will be affected by price. Moreover, the fact that the production process for logs is time-consuming is no real argument for treating it as a fixed endowment at the time of trade, since the same applies not only to virtually all agricultural produce, but also all manufactures as well (for which there may well be long development times prior to production).

If we consider logs not as a natural endowment but as a produced commodity, it becomes clear that the appropriate model will be one of trade in intermediates. Here we run into a problem. Bhagwati (1964) once lamented that the underdevelopment of theories of trade in intermediate goods was one of the central limitations in the pure theory of international trade. Although things have come a long way since then, still relatively little attention is paid to the details of such models. This is perhaps surprising given that most items traded among countries do not represent final consumption items, but rather raw materials, producer goods and processed materials.

Suzuki (1978) classifies the theory of trade in intermediate goods into three categories: (i) the theories examining whether the pattern of comparative advantage and the distribution of income in the model with intermediates are determined by the same conditions as the traditional ones concerning trade in final goods; (ii) the theories investigating whether commercial policy has the same effects; and (iii) the theories regarding the gains from trade in intermediate goods. Unfortunately, none of the trade models that incorporate intermediate goods has really emerged as a benchmark model in the same way as the HOS model is the undisputed workhorse of standard trade theory. Kohli (1991) attributes this to the fact that the models which have been proposed are typically not well suited to empirical implementation due to their intricate production structure (p.62).

The models incorporating intermediates tend to be complex. For this reason a variety of simplifications have been adopted. One is the use of pure intermediates (a good used exclusively as an input into the production process), as in Batra and Casas (1973), as opposed to treating intermediate inputs in the same manner as final goods (such as in Vanek, 1963; Kemp, 1969; and Casas, 1972). Other authors assume that only one industry uses intermediates (Jones, 1971; and Ray, 1975), or that intermediates are used only in fixed proportions (Vanek, 1963; and Batra and Casas, 1973). Still others assume that one or more of the final goods is not produced domestically (Schweinberger, 1975), that the intermediate is exclusively imported (Jones, 1971; and Reidel, 1976), or that the intermediate good is non-traded (Batra and Casas, 1973; and Ray, 1975). Sanyal and Jones (1982) use a model where final goods are non-traded.

With respect to the second category of Suzuki, the vast majority of the literature on intermediate goods and the nature of commercial policy (the primary concern of this study) has been concerned with import restrictions, in particular focusing on the nature and role of effective rates of protection, a concept formally developed by Corden (1966), with further developments in Corden (1969) and (1971). Here we are most interested in the theoretical implications of export restrictions on intermediate goods for domestic processing, and this is an area where there have been relatively few contributions.

Burgess (1976) was one of the first to consider the issue of export restraints and processing in a general equilibrium framework. In his paper a model is specified that explicitly recognises the production of primary products as an activity requiring scarce natural resources in addition to capital and labour. Two final goods are produced along with a primary product, and both use the primary product as an input alongside capital and labour (both of which are fully mobile). The model is perfectly competitive, and all firms operate under constant returns to scale technology. Factors are available in fixed supply. The effects of various processing incentives (export taxes on primary products, import tariffs on final goods, and subsidies to local processing) are examined with respect to the welfare of natural resource owners. Burgess does not consider overall welfare, or the impact of the incentives on the production and trade structure of the economy. Burgess also takes prices as exogenous, and is thus unable to consider the case where the country concerned is a large supplier of either the raw material or the processed good. Burgess (1980) and (1980b) deal with the same issue in the context of a specific-factor model, but again only for a small country.

The first study to consider the export tax explicitly in a two country setting was Suzuki (1978). He presents a model in which three goods are traded between two countries. One of the goods traded is an intermediate good, and the case considered is where the home country exports the intermediate good, and imports a final good which uses that intermediate as an input into its production. He is then able to show that if the home country has monopoly power in its export market, the imposition of an export tax can have the perverse effect of reducing home country welfare through its effect on the home price of the importable. Suzuki does not consider the case where the country exports both the intermediate good, and the good into which it is processed simultaneously.

This issue is taken up by Keppler (1985). The context of the problem considered is where the export restricting country is already an exporter of both the raw and processed goods. The raw product is defined as a pure intermediate good, and the processed product and the importable as pure consumer goods. Production of the importable does not occur within the export restricting country. The model is perfectly competitive and assumes full employment. Limitations of his model include a failure to specify any sort of welfare function, thus making the analysis relevant only for consideration of the distributive effects of the export tax policy.

In a more recent paper, Jones and Spencer (1989) develop a general equilibrium model to consider the issue of further processing and export restrictions. In the paper a two-country, specific-factors model, with fixed coefficients in processing, is developed in a competitive setting to analyse a nation's optimal strategy in restricting raw material exports when allowance is made for potential beneficial effects on the export price of processed goods.

Jones and Spencer build into their model asymmetries such that the raw material is relatively easy to obtain at home than abroad, that all home production of the processed good is destined for export, and that the country exporting both the raw and processed good has already imposed optimal export controls on the raw material in the form of an ad valorem tax, but not on the processed good. By developing their model with the basic supply-side asymmetry that raw material supply elasticity is greater in the home country than abroad they are able to prove that, in the context of their model, if the home country expands its exports of raw materials, world output of processed goods will be enlarged, resulting in a price fall in a stable competitive market. Rephrasing the argument for the analysis of home commercial policy, a contraction in raw materials exports expands processing at home, but contracts processing abroad to a greater extent. The result is a rise in the world price of the processed, which is viewed in beneficial terms at home since the processed good is also exported. The paper then proceeds to analyse welfare effects, and to formulate an expression for optimal control over raw material exports.

Finally, Jones and Spencer (1989) turn the analysis to the optimal response to a foreign tariff on processed exports, in the presence of an existing optimal strategy in controlling exports of the raw material. Perhaps surprisingly, they find that the optimal strategy for dealing with the increase in the foreign tariff may be, rather than to tighten export controls on the raw material even further, to relax the export controls, if the effects on the final goods market are taken into account.

Limitations of the Jones and Spencer study include the use of fixed proportions technology in the production of the processed good, an assumption that may not be appropriate for all processing industries, and the arbitrary assumptions of no domestic consumption of the processed good, and an elasticity of supply of the raw material that is smaller in the foreign country than at home. In the following two sections we develop a model similar to that used by Jones and Spencer, but without any such limiting assumptions. We find that their conclusions do not necessarily continue to hold. We begin with the case of the small economy.

A General Equilibrium Model of a Small Economy

Production

We consider an economy that produces three goods, logs, manufactures and lumber. Logs are produced using labour (L) and the specific factor natural resources (N), while manufactures are produced using labour and the specific factor capital (K). These two industries form what in the parlance of Jones and Spencer (1989) is referred to as the primary tier of the economy. The secondary, or processing, tier consists of lumber production, which is footloose, being produced using labour and logs (the complementarity requirement). The wage rate thus provides the link between competitive costs among the industries. All primary factors of production are assumed to be available in fixed supply, and are internationally immobile. All goods are freely tradeable, and we assume the economy exports logs and lumber in exchange for manufactures.[1] We denote the production of manufactures as industry 0, log production as industry 1, and lumber production as industry 2. We therefore have the following production functions:

$$g_0 = f^0(L_0, K),$$
$$g_1 = f^1(L_1, N),$$
$$g_2 = f^2(L_2, q_1),$$

where g_i is gross output of good i, K is capital, L_i is labour used in sector i, N is natural resources (land), and q_1 is the input of logs into the processing sector (i.e., net output from sector 1, y_1, is given by $g_1 - q_1$, which we assume to be positive). Note that we have assumed for simplicity that the output of the natural resource-based sector is not utilised by the general manufacturing sector. We assume the production functions have the following standard properties: they are positive, continuous, strictly concave for inputs greater than zero, and linearly homogeneous (i.e., we assume that the production technology exhibits constant returns to scale). Assuming the processing sector pays market prices for logs, unit cost functions can then be defined in the following manner:

$$c^0(w,r) \equiv \min_{L_0,K}\{wL_0 + rK : f^0(L_0,K) = 1,\ L_0 \geq 0,\ K \geq 0\},$$
$$c^1(w,n) \equiv \min_{L_1,N}\{wL_1 + nN : f^1(L_1,N) = 1,\ L_1 \geq 0,\ N \geq 0\},$$
$$c^2(w,p_1) \equiv \min_{L_2,q_1}\{wL_2 + p_1q_1 : f^2(L_2,q_1) = 1,\ L_2 \geq 0, q_1 \geq 0\},$$

where w is the factor reward to labour, r is the factor reward to capital, and n is the factor reward to natural resources. The unit cost functions denote the minimum cost of all inputs required to produce one unit of gross output.

With perfect competition ensuring zero profits in all industries, we can now write the profit maximising conditions in terms of the unit cost functions as:

$$c^0(w,r) = p_0, \tag{5.1}$$
$$c^1(w,n) = p_1, \tag{5.2}$$
$$c^2(w,p_1) = p_2, \tag{5.3}$$

where p_i is the price of good i. By Shephard duality, the unit cost functions are positive, linearly homogeneous, and concave in input prices by the properties of the production functions. Furthermore, we know by Shephard's lemma that the partial derivatives of the cost functions with respect to each factor price yield the optimal inputs of that factor per unit of gross output (i.e., the optimal input-output coefficient), which we denote as a_{ij}. Similarly, we know that $a_{2q_1} \equiv c^2_{p_1}(w,p_1) = \partial c^2(w,p_1)/\partial p_1$ is the optimal input of logs per unit of gross output lumber. Note that the optimal usage of labour and logs in producing lumber are functions of both wages and the price of logs, indicating possible substitutability. Assuming full employment of the fixed endowment of factors, we can now define factor market equilibrium by the following set of equations:

$$a_{0K}g_0 = \overline{K}, \tag{5.4}$$

$$a_{0L}g_0 + a_{1L}g_1 + a_{2L}g_2 = \overline{L}, \tag{5.5}$$

$$a_{1N}g_1 = \overline{N}, \tag{5.6}$$

where a bar represents the fixed endowment of factor of production. These equations simply state that total factor usage in the economy is constrained by the fixed initial endowments. Since prices and factor endowments are given, this gives us a set of six equations 5.1-5.6 in six unknowns (w, r, n, g_0, g_1, and g_2). This completes our description of the production side of the small economy.

The GNP Function

In the equations above, we have enough information to consider the impacts of processing incentives on the production side of the economy. However, it will prove useful to our analysis to also aggregate production into the *GNP* function. The *GNP* function provides a very general technology, without requiring us to explicitly deal with the division of the production sector into separate industries. We will make use of the *GNP* function both in describing the budget constraint facing the small economy, and more extensively in the two country model which follows, where the simplifications it allows are most useful. The *GNP* function can be defined as:

$$G(p_0, p_1, p_2, \overline{K}, \overline{L}, \overline{N}) \equiv \max_{g_0, y_1, y_0} \{p_0 g_0 + p_1 y_1 + p_2 g_2 : (g_0, y_1, g_2) \in Y\},$$

where:

$$Y = \{(g_0, y_1, g_2) : g_0 \leq f^0(L_0, K), y_1 \leq f^1(L_1, N) - q_1, g_2 \leq f^2(L_2, q_1), K \leq \overline{K}, L_0 + L_1 + L_2 \leq \overline{L}, N \leq \overline{N}\},$$

is the set of net outputs that can be produced given the endowment vector (i.e., the production possibilities set). The function thus states that the production sector is assumed to choose the vector of net outputs to maximise *GNP*, which is the product of the price vector and net outputs, subject to production possibility set. The *GNP* function can alternatively be expressed in terms of the unit cost functions as:

$$G(p_0, p_1, p_2, \overline{K}, \overline{L}, \overline{N}) \equiv \min_{w, r, n} \{r\overline{K} + w\overline{L} + n\overline{N} : c^0(w, r) \geq p_0, c^1(w, n) \geq p_1, c^2(w, p_1) \geq p_2\},$$

which states that the *GNP* function can be found by finding the factor price vector which minimises the value of factor endowments subject to the constraint that unit cost must not be lower than price for any good.[2]

The *GNP* function can be shown to be positive for all positive prices and factor endowments, continuous, linearly homogeneous and convex in prices for all factor endowments, and non-decreasing and concave in factor endowments for all prices. We further assume for convenience that it is twice continuously differentiable. If this is the case, then by Hotelling's lemma, the partial derivatives of the *GNP* function with respect to prices give the net output supply functions. The partial derivatives with respect to factor endowments give the shadow prices of factors (the marginal increase in *GNP* due to a marginal increase in the factor endowment). In a competitive equilibrium these correspond to market prices for the factors of production.

Demand

We complete our model by specifying the demand side. We make the simplifying assumptions that consumers have identical preferences, and that their individual utility functions can be aggregated into a direct social utility function $\mu(z_0, z_2)$ where z_i is the consumption of good *i*. Note that logs are assumed to be a pure intermediate, and therefore do not enter the utility function directly. They can increase utility only by being processed into lumber, or exported in exchange for manufactures. We assume that the social utility function is non-negative, continuous, quasi-concave, and increasing in consumption of all goods. We further assume that utility does not depend on the quantities of factors supplied (i.e., we maintain our assumption that primary factors are available in perfectly inelastic supply). If we are willing to accept the hypothesis that consumers select the consumption bundle that maximises their utility subject to the constraint imposed by their income, we obtain the indirect utility function:

$$V(p_0, p_2, GNP) \equiv \max_{z_0, z_2} \{ \mu(z_0, z_2) : p_0 z_0 + p_2 z_2 \le GNP, z_0 \ge 0, z_2 \ge 0 \},$$

which indicates the maximum utility attainable with income *GNP* when prices are p_i. The assumptions we have imposed on the social utility function ensure that all income is spent, and therefore the solutions to this problem yield the Marshallian demand functions, $z_0 = D_0(p_0, GNP)$ and $z_2 = D_2(p_2, GNP)$. Minimising the expenditure necessary to attain a target level of utility (u) at given prices then allows us to define the aggregate expenditure function:

$$E(p_0, p_2, u) \equiv \min_{z_0, z_2}\{p_0 z_0 + p_2 z_2 : \mu(z_0, z_2) \geq u\},$$

This function is analogous to the *GNP* function used on the production side, and is non-decreasing, homogeneous of degree one, and concave in prices. We further assume that it is twice continuously differentiable. In view of Shephard's lemma, it can be shown that the partial derivatives of the expenditure function with respect to prices yield the Hicksian demand functions.

The budget constraint facing the economy (the balance of trade) can now be expressed in terms of the *GNP* and expenditure functions as:

$$G(p_0, p_1, p_2, \overline{K}, \overline{L}, \overline{N}) = E(p_0, p_2, u), \tag{5.7}$$

which simply states that the value of total production is equal to the value of total expenditure. Equation 5.7 completes our model for the small economy. We now have seven equations (5.1-5.7), and seven unknowns (w, r, n, g_0, g_1, g_2, and u). This completes our description of the model specification for the small country. We now turn to examining the comparative static characteristics of the specification.

Comparative Statics - Factor Prices

With our model economy specified as above, we can begin to answer a number of relevant questions with respect to the impact of alternative policy options. We start with the effects on factor prices. Since the number of factors is equal to the number of products, in order to determine these we need only consider the profit maximising conditions 5.1-5.3. Differentiating equations 5.1-5.3 totally and expressing in percentage change form yields the following basic matrix equation:

$$\begin{bmatrix} \theta_{0L} & \theta_{0K} & 0 \\ \theta_{1L} & 0 & \theta_{1N} \\ \theta_{2L} & 0 & 0 \end{bmatrix} \begin{bmatrix} \hat{w} \\ \hat{r} \\ \hat{n} \end{bmatrix} = \begin{bmatrix} \hat{p}_0 \\ \hat{p}_1 \\ \hat{p}_2 - \theta_{2q_1}\hat{p}_1 \end{bmatrix}, \tag{5.8}$$

where the θ_{ij} are cost shares, which sum to one, and a hat denotes relative changes (i.e., $\hat{w} \equiv dw/w$). Since we will examine several alternative processing incentives, it is convenient to express the relationships between good and factor prices in this general form.

As our first case, suppose that the economy imposes an export tax on logs with the intention of increasing the level of domestic processing (lumber production). Since world prices are given for the small country, the producer prices are determined by the extent of the export tax t imposed. Assuming a specific duty, we therefore have:

$$p_1 + t = p_1^*, \tag{5.9}$$

where a superscript * is used here to denote the given world price. The export tax thus lowers the price of logs for both lumber and log producers. All other prices remain fixed. For simplicity, we make the standard assumption that the revenue accruing to the government from the policy is redistributed in a lump-sum fashion that does not favour any particular class of factor owner. Making the appropriate adjustments to the matrix equation 5.8 and solving for the relative changes in factor prices yields the following expressions:

$$\hat{w} = (-\theta_{2q_1}/\theta_{2L})\hat{p}_1, \tag{5.10}$$
$$\hat{r} = (\theta_{0L}\theta_{2q_1}/\theta_{2L}\theta_{0K})\hat{p}_1, \tag{5.11}$$
$$\hat{n} = [(\theta_{2L} + \theta_{1L}\theta_{2q_1})/\theta_{2L}\,\theta_{1N}]\hat{p}_1. \tag{5.12}$$

Since the cost shares are all positive, the expressions reveal that an export tax on logs causes the real return to owners of natural resources and owners of capital to unambiguously fall, and the real return to labour to unambiguously rise. The reduction in the price of logs causes an expansion of the processing sector (see the following section for expressions revealing how). The only way to increase production in the model is to acquire more labour, and upward pressure is exerted on the wage rate. Real returns to other factors of production must therefore fall.

Of course, an export tax is not the only means of increasing processing. Perhaps the most obvious alternative is a subsidy to processing. We therefore consider a policy that lowers the price of logs for processing, while holding the price that domestic log producers receive constant. This gives us:

$$p_1^s = p_1 - s, \tag{5.13}$$

where p_1^s is the price paid by producers. Again, for simplicity we make the standard assumption that the subsidy is financed in a non-distortionary manner by lump-sum transfers that do not disfavour any particular factor. Making the appropriate changes to the matrix equation 5.8 and observing its new structure, it is clear that the impact of a processing subsidy on the real wage and the real return to capital will be identical to 5.10 and 5.11, only the change in the real return to natural resources will differ, becoming:

$$\hat{n} = (\theta_{1L}\theta_{2q_1}/\theta_{1N}\theta_{2L})\hat{p}_1^s. \tag{5.14}$$

In other words, the subsidy causes the return to owners of natural resources to fall by less than an export tax that reduces the price to lumber producers by the same amount. This is because the price received by log producers does not fall, while the wage rate rises by the same proportion.

The final policy we consider is an export subsidy on lumber. This has the effect of raising the domestic price of lumber. The solutions for relative changes in factor prices with the assumptions given become:

$$\hat{w} = (1/\theta_{2L})\hat{p}_2, \tag{5.15}$$
$$\hat{r} = (-\theta_{0L}/\theta_{2L}\theta_{0K})\hat{p}_2, \tag{5.16}$$
$$\hat{n} = (-\theta_{1L}/\theta_{2L}\theta_{1N})\hat{p}_2. \tag{5.17}$$

Once again, it is clear that the return to labour rises, and the returns to the other two factors fall. Note also that we can be certain that real wages have in fact risen despite the fact that relative consumer prices have now changed since the cost shares all lie between zero and one, and thus the wage rate rises by a greater proportion than the price of lumber. Labour is thus better off even if it only consumes lumber.

Comparative Statics - Outputs

Also of considerable interest to policy makers are the impacts of processing incentives on outputs and trade, which we describe here. To obtain expressions for the changes in gross output we differentiate the factor market equilibrium equations 5.4-5.6 totally holding factor endowments constant. Expressing in percentage change form we have:

$$\begin{bmatrix} 1 & 0 & 0 \\ \lambda_{0L} & \lambda_{1L} & \lambda_{2L} \\ 0 & 1 & 0 \end{bmatrix} \begin{bmatrix} \hat{g}_0 \\ \hat{g}_1 \\ \hat{g}_2 \end{bmatrix} = \begin{bmatrix} -\hat{a}_{0K} \\ -(\lambda_{0L}\hat{a}_{0L} + \lambda_{1L}\hat{a}_{1L} + \lambda_{2L}\hat{a}_{2L}) \\ -\hat{a}_{1N} \end{bmatrix}, \tag{5.18}$$

where the λ_{ij} are the proportions of factor j used by industry i, and the $\hat{a}_{ij}$ are the proportional changes in the optimal input coefficient of factor j in industry i. We define the elasticity of substitution between labour and capital in industry 0 as:

$$\sigma_0 \equiv (\hat{a}_{0K} - \hat{a}_{0L})/(\hat{w} - \hat{r}). \tag{5.19}$$

Similarly, the elasticities of substitution between labour and natural resources in industry 1, and labour and logs in industry 2 are given by:

$$\sigma_1 \equiv (\hat{a}_{1N} - \hat{a}_{1L})/(\hat{w} - \hat{n}), \tag{5.20}$$
$$\sigma_2 \equiv (\hat{a}_{2q_1} - \hat{a}_{1L})/(\hat{w} - \hat{p}_1), \tag{5.21}$$

respectively. Also, since the distributive share-weighted average of changes in the optimal input-output coefficients along the unit isoquant in each industry must vanish near the cost-minimisation point (i.e., an iso-cost line is tangent to the unit isoquant of each industry), we know that the following set of equations must hold:

$$\theta_L \hat{a}_{0L} + \theta_{0K}\hat{a}_{0K} = 0, \tag{5.22}$$
$$\theta_{1L} \hat{a}_{1L} + \theta_{1N}\hat{a}_{1N} = 0, \tag{5.23}$$
$$\theta_{2L} \hat{a}_{2L} + \theta_{2q_1}\hat{a}_{2q_1} = 0. \tag{5.24}$$

Solving equations 5.19-5.24 for the percentage changes in the optimal input-output coefficients yields:

$$\hat{a}_{0K} = -\theta_{0L}\sigma_0(\hat{r} - \hat{w}), \tag{5.25}$$
$$\hat{a}_{0L} = \theta_{0L}\sigma_0(\hat{r} - \hat{w}), \tag{5.26}$$
$$\hat{a}_{1N} = -\theta_{1L}\sigma_1(\hat{n} - \hat{w}), \tag{5.27}$$
$$\hat{a}_{1L} = \theta_{1N}\sigma_1(\hat{n} - \hat{w}), \tag{5.28}$$
$$\hat{a}_{2q_1} = -\theta_{2L}\sigma_2(\hat{p}_1 - \hat{w}), \tag{5.29}$$
$$\hat{a}_{2L} = \theta_{2q_1}\sigma_2(\hat{p}_1 - \hat{w}). \tag{5.30}$$

These expressions reveal how all three industries economise on the use of labour (which has seen its real return rise). Substituting these expressions back into the differential forms of the factor market equilibrium conditions, and solving for gross outputs, yields after some manipulation:

$$\hat{g}_0 = \theta_{0L}\sigma_0(\hat{r} - \hat{w}), \tag{5.31}$$
$$\hat{g}_1 = \theta_{1L}\sigma_1(\hat{n} - \hat{w}), \tag{5.32}$$
$$\hat{g}_2 = (-\lambda_{0L}/\lambda_{2L})\sigma_0(\hat{r} - \hat{w}) - (\lambda_{1L}/\lambda_{2L})\sigma_1(\hat{n} - \hat{w}) - \theta_{2q_1}\sigma_2(\hat{p}_1 - \hat{w}). \tag{5.33}$$

These expressions reveal the effect of a processing incentives on gross output of manufactures, logs and lumber in terms of the changes in factor prices. Since the cost shares, elasticities of substitution, and factor use proportions are all positive, and we know from above that in the case of all processing incentives the return to natural resources falls and the return to labour rises, the expressions reveal that gross output of manufactures (g_0) and logs (g_1) must fall, and gross output of lumber (g_2) must rise in all cases. Note that the increase in lumber output can be attributed to two effects. The first two terms in 5.33 reveal the effect of labour released from the other two industries, and the third the effect of substitution between labour and logs in the lumber industry. Note that with a processing subsidy, because $\hat{n}$ is smaller, gross output of logs falls by less, and gross output of lumber rises by less (as a consequence of less labour being released from log production). In other words, a subsidy raises processing by a lower proportion than the export tax with an equivalent price effect in this model.

Furthermore, since usage of logs in the lumber industry has risen ($\hat{a}_{2q_1} = -\theta_{2L}\sigma_2(\hat{p}_1 - \hat{w})$) and gross output of logs has fallen, it must be the case that net output of logs has also fallen. Given that logs are not consumed directly, this means that exports of logs (which are therefore equal to net output) fall with the use of all processing incentives, as expected. In the case of the export tax and the processing subsidy, consumer prices remain unchanged, but real income falls. Consumption of both final goods must therefore fall. The implication is that exports of lumber must rise, again as one would expect. The effect on imports of manufactures is less clear. One might assume that since real income has fallen the policies would reduce imports, in line with what standard trade theory would lead us to expect (in particular in the case of an export tax). However, the processing incentives also draw labour out of production of the importable, imports may in fact therefore rise or fall with the use of processing incentives.

Welfare Effects

We now proceed to consider the implications of processing incentives for overall welfare for the small country. We begin with an export tax. With an export tax in place, total revenue equals the value of production plus the revenue from the export tax. We thus rewrite the budget constraint 5.7 as:

$$G(p_0, p_1, p_2, \overline{K}, \overline{L}, \overline{N}) + ty_1 = E(p_0, p_2, u),$$

where y_1 is net output (exports) of logs. By Hotelling's lemma the partial derivatives of the *GNP* function with respect to prices yield the optimal net outputs. Therefore differentiating the budget constraint totally holding factor endowments and all prices except p_1 constant gives us:

$$E_u du = y_1 dp_1 + tdy_1 + y_1 dt.$$

Making use of the fact that by differentiating equation 5.9 totally we have $dp_1 = dp_1^* - dt$, and the fact that the small country assumption implies the world price of logs is unchanged, this reduces to simply:

$$E_u du = tdy_1. \qquad (5.34)$$

As we know from equations 5.32 and 5.33, dy_1 is negative, and therefore social welfare must fall with the imposition of an export tax on logs, as expected. The welfare impacts of the other processing incentives can be derived in a similar manner. For a processing subsidy we have:

$$E_u du = -sdq_1, \qquad (5.35)$$

where q_1 is the amount of logs used in lumber production. Since we know from above that q_1 rises, the expression shows that welfare falls, but by a lesser amount than with an export tax of the same magnitude, as expected. This can be demonstrated by noting that from the definition of net output of logs we have $dy_1 = dg_1 - dq_1$, and we know from our results above that $dg_1 < 0$ and $dq_1 > 0$. With an export subsidy we have:

$$E_u du = -s dx_2, \tag{5.36}$$

where s is the per unit subsidy and $x_2 = g_2 - z_2$ is the volume of lumber exports. Once again, it is clear that social welfare falls. All forms of processing incentives result in lower social welfare. Note also that the optimal export tax follows immediately from 5.34, setting the marginal condition du=0, the optimal tax is clearly t=0.[3] In other words, for the small country no form of processing incentive can raise welfare, a general result that is well-known.

Foreign Ownership

What happens if we allow for foreign ownership? In this section we deal in a simple way with the very current issue of foreign-owned factors of production, which until now has been sidelined in our analysis. This is a complex issue, and we deal here only with certain static aspects.[4] With foreign direct investment, the income paid for the services of foreign owned factors is not part of domestic income. One of the stylised facts of foreign ownership as it applies to the forestry industry of New Zealand, is that foreigners have bought a large proportion of the forest resource, but have made only very small investments in downstream processing. To reflect this stylised fact in our model, we assume for simplicity that the entire specific factor (and no other factors) used in log production (N) is foreign-owned. In terms of the welfare results we obtain, this assumption means that we are dealing here with the 'best case' scenario. With an export tax on logs, the budget constraint the economy faces is given by:

$$G(p_0, p_1, p_2, \overline{K}, \overline{L}, \overline{N}) + ty_1 - n\overline{N} = E(p_0, p_2, u),$$

where $n\overline{N}$ is the payment for the use of foreign factors of production. Again, we derive the welfare effects of an export tax on logs by differentiating the new budget constraint totally holding factor endowments and all prices except p_1 constant. Simplifying the resulting equation as above yields:

$$E_u du = t dy_1 - \overline{N} dn. \tag{5.37}$$

This shows that the effect of the export tax for the small country on overall welfare can be decomposed into two components, a volume of trade effect (which is negative) and an income transfer effect. Since we know from 5.12 that *dn* is negative, it is clear that a tax could result in an increase in domestic welfare, even though the country is small, provided the welfare gain from the income transfer exceeds the welfare loss from reduced exports. The 'optimal' export tax will be where the incremental gain from the income transfer exactly equals the incremental deadweight loss.

Thus the argument that we should restrict exports of logs and encourage domestic processing because foreign owners are chopping down our forests and exporting them as logs has some genuine intellectual as well as emotive appeal. A processing or export subsidy will have a similar effect for a small country, since both lower the return to owners of natural resource (and all, in effect, restrict exports).[5] Of course, a much less distorting measure would be to simply tax foreign factors directly, but an export tax may be more politically viable since it could be packaged as a policy aimed at increasing processing, rather than an attempt to transfer income from foreign owned factors. Whether one considers this a good argument will depend on one's opinion of nationalisation of foreign assets, since this is essentially what such a policy amounts to (although one could easily argue that this is the case for all forms of 'optimal' tariff arguments).

Escalating Foreign Tariffs

Let us suppose that there exists a country (which we will call foreign) that is a sufficiently large importer of lumber and logs that it can influence the world prices of these goods (assume that all other exporting countries are also small). Suppose that this country decided to impose an import tariff on lumber. By exercising its monopsony power, it would be able to drive down the price of lumber on world markets, and thus make itself better off. How would such a policy affect our home country in this model? If we consider the impact of the policy on home factor prices, it is clear that the relevant equations are 5.15-5.17, the same as those that determined the effect of an export subsidy. However, in this case the lumber price has fallen, therefore it must be the case that the return to labour falls, and the returns to capital and natural resources rise (overall welfare must, of course, decline). The effects on output are given by 5.31-5.33, and it is clear that output of logs and manufactures must rise, while output of lumber must fall.

What is the optimal policy response for the home country in this case? In the absence of foreign ownership of factors of production, in the case of the small economy no form of retaliation is rational. If it cannot affect world prices, it cannot affect foreign processors, and it will only lower its own welfare further by responding. The home country can thus do nothing.

But what if we have foreign ownership? Let us imagine a situation where the foreign country that is imposing the escalating tariff is the same as the country from which the foreign investment in natural resources in the home country originates (the parallel with Japan is suggestive). By equation 5.37, a decrease in the home lumber price must result in increased transfers from the home to the foreign country. In other words, the existence of foreign investment in the home country provides an increased incentive on the part of the foreign country to impose an escalating tariff. The foreign country gains not only through the terms of trade effect, but also through the increased transfers. In this case, however, the home country is not powerless in the face of such escalating tariffs. As we have seen, the imposition of an export tax on logs (or a direct tax on the foreign transfers) provides a means of grabbing back the income stream. This is an intriguing possibility, and one that we have not seen mentioned in the theoretical literature.

General Equilibrium for a Trading World

We have now completed our formal analysis of the impact of processing incentives for the small economy. Once again, we emphasise that we believe that for an economy like New Zealand, the small country analysis is extremely important. However, it is only able to tell a part of the story. Not all countries are small, and certain questions, in particular with respect to the notion of feedback, cannot be addressed in the context of exogenous prices. We therefore now turn to a model in which we have two countries (representing New Zealand and the rest of the world). This will enable us to observe the effect of processing incentives in a true general equilibrium. In this section we will work with an aggregated model (i.e., in terms of the *GNP* and expenditure functions). Our objective is essentially to determine the effect of processing incentives on the world price vector. Once this is determined (if this is possible), the more detailed underlying model can be consulted to infer the effects on factor prices, outputs and social welfare.

Net Export Functions

Using the *GNP* and expenditure functions defined above, we can define the net revenue (or net balance of payments) function for the home country as:

$$S(p_0, p_1, p_2, \overline{K}, \overline{L}, \overline{N}, u) \equiv G(p_0, p_1, p_2, \overline{K}, \overline{L}, \overline{N}) - E(p_0, p_2, u).$$

This is merely a new way of expressing 5.7. The net revenue function is linearly homogeneous and convex in prices, and decreasing in *u*. It also follows by the derivative properties of the *GNP* and expenditure functions that:

$$S_0(p_0, p_1, p_2, \overline{K}, \overline{L}, \overline{N}, u) = G_0(p_0, p_1, p_2, \overline{K}, \overline{L}, \overline{N}) - E_0(p_0, p_2, u), \quad (5.38)$$
$$S_1(p_0, p_1, p_2, \overline{K}, \overline{L}, \overline{N}) = G_1(p_0, p_1, p_2, \overline{K}, \overline{L}, \overline{N}), \quad (5.39)$$
$$S_2(p_0, p_1, p_2, \overline{K}, \overline{L}, \overline{N}, u) = G_2(p_0, p_1, p_2, \overline{K}, \overline{L}, \overline{N}) - E_2(p_0, p_2, u). \quad (5.40)$$

These functions define Hicksian net exports (or excess supplies). Recalling that by Hotelling's lemma the partial derivative of the *GNP* function with respect to prices gives the net output supply functions, it is clear that the foregoing equations simply state that the Hicksian net export function is equal to the difference between net output of a good, and domestic Hicksian demand. Note that the utility level is not an argument in equation 5.39 by the assumption that logs are not directly consumed.

The Model

Letting the foreign country be denoted by a superscript *, we can now describe the equilibrium between the two countries. With free trade between the two countries, there will only be one set of prices, and we choose to use the prices of the home country to denote these (Dixit and Norman, 1980, refer to this as the 'imperial convention'). Equilibrium is then described by the following set of equations:[6]

$$S(p_0, p_1, p_2, \overline{K}, \overline{L}, \overline{N}, u) = 0,$$
$$S^*(p_0, p_1, p_2, \overline{K}^*, \overline{L}^*, \overline{N}^*, u^*) = 0,$$
$$S_0(p_0, p_1, p_2, \overline{K}, \overline{L}, \overline{N}, u) + S_0^*(p_0, p_1, p_2, \overline{K}^*, \overline{L}^*, \overline{N}^*, u^*) = 0,$$
$$S_1(p_0, p_1, p_2, \overline{K}, \overline{L}, \overline{N}) + S_1^*(p_0, p_1, p_2, \overline{K}^*, \overline{L}^*, \overline{N}^*) = 0,$$

$$S_2(p_0, p_1, p_2, \overline{K}, \overline{L}, \overline{N}, u) + S_2^*(p_0, p_1, p_2, \overline{K}^*, \overline{L}^*, \overline{N}^*, u^*) = 0.$$

This is a set of five equations. However, by Walras' law, if M-1 markets are in equilibrium the last market is also in equilibrium. Hence one equation is redundant. The equilibrium prices are, however, only determined up to a factor of proportionality, and some normalisation of prices is required. We choose to follow the procedure used by Dixit and Norman (1980) and select a common normalisation for both countries. The most simple normalisation procedure is to let the price of manufactured goods play the role of numeraire, and the obvious choice of the market clearing condition to drop is then that for manufactured goods. The equilibrium conditions then become:

$$S(p_1, p_2, \overline{K}, \overline{L}, \overline{N}, u) = 0, \tag{5.41}$$

$$S^*(p_1, p_2, \overline{K}^*, \overline{L}^*, \overline{N}^*, u^*) = 0, \tag{5.42}$$

$$S_1(p_1, p_2, \overline{K}, \overline{L}, \overline{N}) + S_1^*(p_1, p_2, \overline{K}^*, \overline{L}^*, \overline{N}^*) = 0, \tag{5.43}$$

$$S_2(p_1, p_2, \overline{K}, \overline{L}, \overline{N}, u) + S_2^*(p_1, p_2, \overline{K}^*, \overline{L}^*, \overline{N}^*, u^*) = 0. \tag{5.44}$$

where the argument of manufactured good price (which is now equal to one) has been dropped for convenience. There are now four equations in the two utility levels and two relative prices. Note, however, that given prices, the utility levels of home and foreign can be determined from the budget constraints that these countries face, 5.41 and 5.42, in the same way as in the small country case. The prices can be determined from equations 5.43 and 5.44, but will be functions of the levels of utility. It would be preferable if we could eliminate the utility levels from equation 5.44. This can be facilitated by the use of the Marshallian net export supply functions, which are obtained by substituting the indirect utility function for the home country (and the foreign equivalent) into equation 5.44 to obtain:

$$X_2(p_1, p_2, \overline{K}, \overline{L}, \overline{N}) + X_2^*(p_1, p_2, \overline{K}^*, \overline{L}^*, \overline{N}^*) = 0, \tag{5.45}$$

where X_2 is the home Marshallian net export supply of lumber, defined as:

$$\begin{aligned} X_2(p_1,p_2,\overline{K},\overline{L},\overline{N}) &= G_2(p_1,p_2,\overline{K},\overline{L},\overline{N}) - E_2(p_2,V(p_2,G(p_1,p_2,\overline{K},\overline{L},\overline{N}))) \\ &= G_2(p_1,p_2,\overline{K},\overline{L},\overline{N}) - D_2(p_2,G(p_1,p_2,\overline{K},\overline{L},\overline{N})), \end{aligned}$$

where D_2 is the Marshallian home demand for logs (see Woodland, 1982, pp.152-4 for details on the equivalence of the two expressions). Similar expressions can be derived for the foreign Marshallian net export supply of lumber. We re-express equation 5.43 as the following to maintain consistency in the notation:

$$X_1(p_1, p_2, \overline{K}, \overline{L}, \overline{N}) + X_1^*(p_1, p_2, \overline{K}^*, \overline{L}^*, \overline{N}^*) = 0, \qquad (5.46)$$

This is a set of two equations (5.45 and 5.46) in two unknowns (the two relative prices). The equations simply state that prices of both logs and lumber adjust so that world excess supplies of the final product and the intermediate good are equal to zero at equilibrium. This completes our description of the model.[7]

Comparative Statics – An Increase in the Natural Resource

In order to develop an understanding of the way in which this model works, it is useful to devote some time to examining a simple comparative statics exercise in the absence of trade taxes. An interesting question is what happens when the home endowment of natural resources increases. (Let us imagine that land previously thought unsuitable for growing trees is suddenly found to be suitable after all, effectively resulting in an increase in the natural resource endowment). How would this discovery affect prices of logs and lumber? In fact, as will become clear, there is no clear-cut answer to this question, but it serves as a useful example of the problems involved in analysing flows of interrelated goods like logs and lumber. We begin by totally differentiating equations 5.45 and 5.46, holding the endowments of all other factors and the natural resource endowment of the foreign county constant. Expressing the resulting equations in matrix form we have:

$$\begin{bmatrix} (X_{11} + X_{11}^*) & (X_{12} + X_{12}^*) \\ (X_{21} + X_{21}^*) & (X_{22} + X_{22}^*) \end{bmatrix} \begin{bmatrix} dp_1 \\ dp_2 \end{bmatrix} = \begin{bmatrix} -X_{1\overline{N}} \\ -X_{2\overline{N}} \end{bmatrix} d\overline{N}, \qquad (5.47)$$

where a second numerical subscript denotes the partial derivative with respect to the price of good *i* (e.g., $X_{11} \equiv \partial X_1 / \partial p_1$), etc. We re-express 5.47 for convenience as:

$$\delta \begin{bmatrix} dp_1 \\ dp_2 \end{bmatrix} = \begin{bmatrix} -X_{1\overline{N}} \\ -X_{2\overline{N}} \end{bmatrix} d\overline{N},$$

where $|\delta| = (X_{11} + X_{11}^*)(X_{22} + X_{22}^*) - (X_{12} + X_{12}^*)(X_{21} + X_{21}^*)$. In order to determine the price effect the factor augmentation, we need to determine first the sign of this determinant. Using the definition of the home Marshallian net export function for lumber, its foreign equivalent, and the corresponding equations for the log market it can be shown that the following definitions hold:[8]

$$X_{22} = S_{22} - D_{2I} X_2, \tag{5.48}$$
$$X_{21} = S_{21} - D_{2I} X_1, \tag{5.49}$$
$$X_{22}^* = S_{22}^* - D_{2I}^* X_2^*, \tag{5.50}$$
$$X_{21}^* = S_{21}^* - D_{2I}^* X_1^*, \tag{5.51}$$
$$X_{11} = S_{11} = G_{11}, \tag{5.52}$$
$$X_{12} = S_{12} = G_{12}, \tag{5.53}$$
$$X_{11}^* = S_{11}^* = G_{11}^*, \tag{5.54}$$
$$X_{12}^* = S_{12}^* = G_{12}^*, \tag{5.55}$$

where a subscript *I* is used to denote a partial derivative with respect to income (which is equal to *GNP*). The first four expressions denote the own and cross price derivatives of the Marshallian net export function of lumber for the home and foreign country, respectively. Consider expression 5.48. This shows that the effect of a change in the price of lumber can be decomposed into two parts. The first part, S_{22}, is the total substitution effect of the price change (the sum of substitution along the production possibility frontier and by all consumers along the their indifference curves). The second part, $D_{2I} X_2$, is the income effect, where a price increase (decrease) raises (lowers) *GNP*, and thus increases (decreases) consumption of lumber. Expression 5.50, the foreign equivalent, can be interpreted similarly.

Now let us consider expression 5.49. This shows the effect of a change in the price of logs on net exports of lumber. Once again, the effect can be decomposed into two parts. The first part, S_{21}, illustrates the effect changes in log prices has on production of lumber. The second part, $D_{2I} X_1$, is the impact on lumber consumption of the change in income which a change in log prices implies. Expression 5.51 can be interpreted in the same way.

Finally, consider the last four expressions, 5.52-5.55. These follow immediately from the fact that for logs (which are not consumed) the Marshallian net export supply is equivalent to the Hicksian net export supply, and are the own and cross price derivatives of the net export functions of logs. Note that with changes in either log or lumber prices, we observe only substitution along the production possibility frontier. There is no income effect because this is a pure intermediate.

Substituting expressions 5.48-5.55 into the definition for the determinant $|\delta|$, we obtain:

$$|\delta| = (S_{11} + S_{11}^{*})(S_{22} + S_{22}^{*} - D_{2I}X_2 - D_{2I}^{*}X_2^{*}) \\ -(S_{12} + S_{12}^{*})(S_{21} + S_{21}^{*} - D_{2I}X_1 - D_{2I}^{*}X_1^{*}),$$

which, by the fact that in equilibrium foreign exports must be the negative of home exports, simplifies to:

$$|\delta| = (S_{11} + S_{11}^{*})\left[S_{22} + S_{22}^{*} + (D_{2I} - D_{2I}^{*})X_2^{*}\right] \\ -(S_{12} + S_{12}^{*})\left[S_{21} + S_{21}^{*} + (D_{2I} - D_{2I}^{*})X_1^{*}\right]. \qquad (5.56)$$

Given this expression, it follows that $|\delta|$ will be positive if (1) $D_{2I} = D_{2I}^{*}$, and (2) both S and S^{*} are strictly convex in p_1 and p_2. The first condition is that each country has the same marginal propensity to consume lumber.[9] Condition (2) guarantees that $(S_{11} + S_{11}^{*})$ and $(S_{22} + S_{22}^{*})$ are positive, and that $(S_{11} + S_{11}^{*})(S_{22} + S_{22}^{*}) > (S_{12} + S_{12}^{*})(S_{21} + S_{21}^{*})$. We assume that both conditions hold, and that $|\delta|$ is therefore positive.

The effects on prices of logs and lumber are given by the following two equations:

$$\frac{dp_1}{d\overline{N}} = \frac{1}{|\delta|}\begin{vmatrix} -X_{1\overline{N}} & (X_{12} + X_{12}^{*}) \\ -X_{2\overline{N}} & (X_{22} + X_{22}^{*}) \end{vmatrix} = \frac{|\alpha|}{|\delta|}, \qquad (5.57)$$

where $|\alpha| = -X_{1\overline{N}}(X_{22} + X_{22}^{*}) + X_{2\overline{N}}(X_{12} + X_{12}^{*})$, and,

$$\frac{dp_2}{d\overline{N}} = \frac{1}{|\delta|}\begin{vmatrix} (X_{11} + X_{11}^{*}) & -X_{1\overline{N}} \\ (X_{21} + X_{21}^{*}) & -X_{2\overline{N}} \end{vmatrix} = \frac{|\beta|}{|\delta|}, \qquad (5.58)$$

where $|\beta| = -X_{2\bar{N}}(X_{11} + X_{11}^{*}) + X_{1\bar{N}}(X_{21} + X_{21}^{*})$. Since we have already determined that $|\delta|$ is positive, the key to determining the price effect of natural resource augmentation lies in determining the signs of $|\alpha|$ and $|\beta|$. We begin with $|\alpha|$. We can decompose this as follows:

$$|\alpha| = -G_{1\bar{N}}(S_{22} + S_{22}^{*}) + (G_{2\bar{N}} - D_{2I}n)(S_{12} + S_{12}^{*}), \tag{5.59}$$

where use is made of the fact that $D_{2I} - D_{2I}^{*} = 0$ by the stability condition, and that the partial derivative of the *GNP* function with respect to a factor gives the shadow price (which in perfect competition is equal to the factor reward) of that factor (i.e., $G_{\bar{N}} \equiv \partial G / \partial N = n$). $G_{1\bar{N}}$ is the home country response of net output (exports) of logs to an increase in natural resources, and is positive. The factor $(S_{22} + S_{22}^{*})$ is also positive, as discussed above. Hence the first term in the expression is negative. How can this term be interpreted? The answer is that the increase in the natural resource endowment increases exports of logs, which tends to push the price of logs downward. Turning now to the second term in the expression. The first factor of this term, $(G_{2\bar{N}} - D_{2I}n)$, decomposes the change in lumber exports into two components, a production and consumption effect. The first term, $G_{2\bar{N}}$, is the home country response of lumber production to an increase in natural resources, which is negative (as labour is drawn into log production). The second part of this factor is positive so long as lumber is a normal good, hence this factor is negative. The second factor is this term $(S_{12} + S_{12}^{*})$ is also negative. It measures the change in net log output with an increase in the price of lumber (which was shown to be negative is the analysis above) in the home and foreign countries. Hence the second term is positive. This can be interpreted as the effect on log prices of reduced trade in lumber. The reduction in lumber trade tends to pull up the price of logs. Overall therefore, the change in the log price in indeterminate, it will depend on which of the two influences on log price dominates. If the change in lumber production in response to an increase in the natural resource endowment is small (this will be the case the amount of labour used in lumber production is large relative to the amount used in log production)[10] and the marginal propensity to consume lumber is also small, then the second term in 5.59 is small, and we expect to see the factor augmentation result in lower log prices.

The expression for the effect on the price of lumber tells a similar story. We can decompose this as follows:

$$|\beta| = -(G_{2\overline{N}} - D_{2l}n)(S_{11} + S^*_{11}) + G_{1\overline{N}}(S_{21} + S^*_{21}). \qquad (5.60)$$

The first term in the expression will be positive, and this represents the effect of reduced lumber exports (which tends to pull the lumber price up). The second term will be negative (since $(S_{21} + S^*_{21})$ is negative, an increase in the log price lowers exports of lumber), and this represents the effect of increased log exports (which tends to pull lumber prices down). Once again, we cannot sign this expression in general.

It may seem that we have not learnt a great deal from this exercise, since the comparative statics cannot be signed. This is not true, however. The example clearly reveals the fundamental nature of the problem of dealing with interrelated goods such as logs and lumber, an issue which we expand on by introducing trade taxes in the following section.

Price Effects of Trade Taxes

We now formally introduce processing incentives into the model. We consider the implications of the home country imposing an export tax on logs, and the foreign country imposing an import tariff on lumber. With trade taxes the prices of logs and lumber in the home and foreign markets will differ to the extent that trade taxes are used. We denote the home export tax on logs by t_1 and the foreign import tariff on lumber by t^*_2. The budget constraints faced by the home and foreign countries must now be modified to take into account the revenue that accrues to the government from trade taxes. They thus become:

$$S(p_1, p_2, \overline{K}, \overline{L}, \overline{N}, u) + t_1 G_1(p_1, p_2, \overline{K}, \overline{L}, \overline{N}) = 0,$$
$$S^*(p^*_1, p^*_2, \overline{K}^*, \overline{L}^*, \overline{N}^*, u^*) + t^*_2 S_2(p^*_1, p^*_2, \overline{K}^*, \overline{L}^*, \overline{N}^*, u^*) = 0,$$

where $p_1 + t_1 = p^*_1$ and $p_2 + t^*_2 = p^*_2$. From these we derive, in the same manner as above, the following equilibrium conditions in terms of the Marshallian net export functions:

$$X_1(p_1, p_2, t_1, \overline{K}, \overline{L}, \overline{N}) + X^*_1(p^*_1, p^*_2, t^*_2, \overline{K}^*, \overline{L}^*, \overline{N}^*) = 0, \qquad (5.61)$$
$$X_2(p_1, p_2, t_1, \overline{K}, \overline{L}, \overline{N}) + X^*_2(p^*_1, p^*_2, t^*_2, \overline{K}^*, \overline{L}^*, \overline{N}^*) = 0, \qquad (5.62)$$

This is a system of four equations (including the price relationships) in four unknowns. To illustrate the nature of the problem involved in determining the price vector, we consider the impact of a home export tax on logs, starting from free trade (which simplifies the analysis somewhat by making the initial tax revenue equal to zero). Differentiating the equilibrium conditions totally for a small change in the export tax from zero, holding the factor endowments of both countries constant, and the foreign tariff on lumber constant at zero we have:

$$X_{11}dp_1 + (X_{12} + X_{12}^*)dp_2^* + X_{11}^*dp_1^* = 0,$$
$$X_{21}dp_1 + (X_{22} + X_{22}^*)dp_2^* + X_{21}^*dp_1^* = 0,$$

where use is made of the fact that, in the absence of trade taxes on lumber, $dp_2 = dp_2^*$. If we also make use of the fact that $dp_1 = dp_1^* - dt_1$, and express in matrix form, this reduces to:

$$\begin{bmatrix} (X_{11} + X_{11}^*) & (X_{12} + X_{12}^*) \\ (X_{21} + X_{21}^*) & (X_{22} + X_{22}^*) \end{bmatrix} \begin{bmatrix} dp_1^* \\ dp_2^* \end{bmatrix} = \begin{bmatrix} X_{11} \\ X_{21} \end{bmatrix} dt_1, \tag{5.63}$$

which can be expressed as:

$$\delta \begin{bmatrix} dp_1^* \\ dp_2^* \end{bmatrix} = \begin{bmatrix} X_{11} \\ X_{21} \end{bmatrix} dt_1,$$

where δ is as defined above, and $|\delta|$ is therefore positive with the stability condition imposed. The solutions for the changes in foreign prices when the export tax is imposed are then given by:

$$\frac{dp_1^*}{dt_1} = \frac{1}{|\delta|} \begin{vmatrix} X_{11} & (X_{12} + X_{12}^*) \\ X_{21} & (X_{22} + X_{22}^*) \end{vmatrix} = \frac{|\phi|}{|\delta|}, \tag{5.64}$$

and,

$$\frac{dp_2^*}{dt_1} = \frac{1}{|\delta|} \begin{vmatrix} (X_{11} + X_{11}^*) & X_{11} \\ (X_{21} + X_{21}^*) & X_{21} \end{vmatrix} = \frac{|\varphi|}{|\delta|}, \tag{5.65}$$

respectively. As in the case of the factor augmentation, determining the signs of the solution depends on determining the signs of $|\phi|$ and $|\varphi|$. We begin with $|\phi|$. Following the same techniques as we used in the example of factor augmentation, this can be decomposed as:

$$|\phi| = S_{11}(S_{22} + S^{*}_{22}) - (S_{21} - D_{2l}X_1)(S_{12} + S^{*}_{12}). \tag{5.66}$$

The corresponding decomposition of $|\varphi|$ is given by:

$$|\varphi| = (S_{11} + S^{*}_{11})(S_{21} - D_{2l}X_1) - S_{11}(S_{21} + S^{*}_{21}). \tag{5.67}$$

Neither of these expressions can be signed unambiguously, a result which should not be too surprising given the analysis of factor augmentation considered above. Both terms in expression 5.66 are positive, while both terms in expression 5.67 are negative.

In order to illustrate what is going on, it is perhaps useful to revisit our spatial equilibrium diagram from Chapter 4, and the example of an export tax on logs in a special case with fixed proportions technology, holding all other prices and output constant. Consider again Figure 4.3. Letting home be country A, and foreign be country B, logs (*R*) and lumber (*L*). The spatial equilibrium is determined by the price at which the excess supply of lumber from country *A* (*XSL*) is equal to the excess demand for lumber in region *B* (*XDL*). This occurs at price *PL*. The physical requirements for lumber production in each country thus determined, the excess supply and demand for logs in each country follows, and equilibrium is determined in the log market as well. Given that logs are the major input into lumber, changes in the price of logs should impact on the supply of lumber. An export tax on logs lowers the price of logs in the exporting country and raises the price in the importing country (*XSR* shifts up). This is effectively a subsidy to lumber production in the exporting country, and a tax on lumber production in the importing country. We would expect the tax to result therefore in a contraction of supply in the importing country. This results in an upward shift in the excess demand for lumber. This is the feedback effect. Note that the shift in supply functions feeds back into the demand for logs. A new equilibrium will eventually be reached where the price paid for logs is higher in the importing than in the exporting country (although it may or may not be higher than prior to the imposition of the tax). The restriction of exports of logs unambiguously causes an expansion of lumber exports if there is feedback. However, note also that the tax also causes the excess supply curve in the exporting country to expand, so the effect on lumber prices is ambiguous, in the diagram it is unchanged.[11]

This partial equilibrium diagrammatical treatment is analogous to our general equilibrium mathematical treatment. Notice that if we assume that the effect of a rise in log prices on the price of lumber in the home country is small, and that lumber consumption is only a small proportion of total consumption (and hence $(S_{21} - D_{2I}X_1)$ is small), expressions 5.66 and 5.67 reveal that an export tax will tend to raise the foreign price of both logs and lumber (a situation that would be viewed favourably by a country exporting both of these products). It is important to note however, that it need not do so. Thus the assumption of a positive feedback effect in prices is not justified. Wiseman and Sedjo (1981) treat the limiting case of feedback as a rise in the price of logs having no effect on the world price of lumber. In fact, a rise in the price of logs could conceivably be accompanied by a decline in the price of lumber.

It is useful here to contrast our theoretical results with those obtained by Jones and Spencer (1989). There are two main differences between the model used here and that used by Jones and Spencer. The first is that they assume fixed proportions technology in the production of lumber for the home country. The second is that they do not allow for the consumption of lumber in the home country. They then show that an export tax raises the price of logs and lumber. However, 5.66 and 5.67 reveal that there is no general reason to expect this to be the case.

From the total differential form of the budget constraint faced by each country, it follows that the welfare effect of a small home export tax on logs starting from free trade will be given by the following equations:

$$E_u du = X_1 dp_1^* + X_2 dp_2^*, \tag{5.68}$$

$$E_{u*} du^* = X_1^* dp_1^* + X_2^* dp_2^* = -X_1 dp_1^* - X_2 dp_2^*, \tag{5.69}$$

which show that the effect of the export tax will be of the opposite sign for the home and the foreign country. However, since the price effects are ambiguous, there is little that we can say about the welfare effects, or optimal strategies in general, unless we make some further restricting assumptions. One possibility which is particularly relevant for the New Zealand case is that the home country is a 'large' supplier of the raw material, but is insignificant with respect to the processed market. In this case we can say something about the optimal export tax on logs. The indirect trade utility function for the home country can be defined as:

$$H(p_1, p_2, \overline{K}, \overline{L}, \overline{N}) \equiv \max_{z_0, z_2} \{\mu(z_0, z_2) : p_0 z_0 + p_2 z_2 \le GNP, z_0 \ge 0, z_2 \ge 0\}$$
$$\equiv V(p_0, p_2, GNP)$$
$$\equiv \{u : S(p_1, p_2, \overline{K}, \overline{L}, \overline{N}, u) + t_1 G_1(p_1, p_2, \overline{K}, \overline{L}, \overline{N})\}.$$

The foreign indirect utility function can be defined similarly. The functions are quasi-convex in prices, and describe the set of home and foreign (respectively) utility levels at which the budget constraint holds.[12] We can therefore determine the export tax on logs that maximises home utility (i.e., the optimal export tax) using this function. Since we assume that the home country is small with respect to the processed good market, but large in the log market, the policy problem for the home country is to choose t_1 (i.e., the price of logs) so as to maximise its national welfare, subject to its budget constraint, and the market clearing condition for logs. From 5.61, the domestic price of logs is some function of the foreign price of logs. Using the implicit function theorem, it follows that $p_{1P} = \partial p_1 / \partial p_1^* = -X_{11}^* / X_{11}$, which is negative since both X_{11} and X_{11}^* are positive. That is to say, in this case the export tax will raise the foreign relative price of logs, and lower the home relative price of logs (driving a wedge between the two). The maximisation problem is:

$$\max_{p_1^*} \{H(p_1, p_2, \overline{K}, \overline{L}, \overline{N})\},$$

and the first order condition for maximisation is therefore:

$$H_P = \frac{\partial H}{\partial p_1^*} = 0.$$

Rewriting the home budget constraint we have:

$$Z = Z(H(p_1, p_2, \overline{K}, \overline{L}, \overline{N}), p_1, p_2, \overline{K}, \overline{L}, \overline{N}) = 0,$$
$$= S(p_1, p_2, \overline{K}, \overline{L}, \overline{N}, u) + (p_1^* - p_1) G_1(p_1, p_2, \overline{K}, \overline{L}, \overline{N}) = 0.$$

By the implicit function theorem, it follows that:

$$
\begin{aligned}
H_P &= (\partial Z/\partial p_1^*)/(\partial Z/\partial H) \\
&= \frac{1}{E_u}[S_1 p_{1P} + G_1 + (p_1^* - p_1)G_{11}p_{1P} - G_1 p_{1P}] \\
&= \frac{1}{E_u}\left[G_1 + (p_1^* - p_1)G_{11}\left(-\frac{X_{11}^*}{X_{11}}\right)\right] \\
&= \frac{1}{E_u}[G_1 - t_1 X_{11}^*],
\end{aligned} \tag{5.70}
$$

where use is made of the fact that the home price is a function of the foreign price, and the fact that, in the absence of final consumption of logs, the functions S_1, G_1 and X_1 are the same thing (i.e., they all represent net exports of logs). Solving the first order condition for a maximum for the optimal export tax (in the form of a specific duty) then yields:

$$
t_1 = G_1/X_{11}^*. \tag{5.71}
$$

We can express this in another form that is better known by noting that by definition $G_1 = -G_1^*$, and that we obtain the ad-valorem export tax (which we denote as T_1) by dividing through by the foreign price to obtain:

$$
T_1 = \frac{t_1}{p_1^*} = -\frac{G_1^*}{p_1^*}\cdot\frac{1}{X_{11}^*} = \frac{1}{\eta^*}, \tag{5.72}
$$

where η^* is the foreign import demand elasticity for logs. In words, in this rather special case, the optimal ad-valorem export tax is given by the inverse of the foreign import demand elasticity for logs, a familiar result.

Foreign Tariffs

To illustrate the general problem of determining the price vector further, we consider as a final example the effect of a change in the foreign import tariff on lumber (starting from a zero tariff). The solutions for the changes in foreign prices are given by:

$$
\frac{dp_1^*}{dt_2^*} = \frac{1}{|\delta|}\begin{vmatrix} X_{12} & (X_{12} + X_{12}^*) \\ X_{22} & (X_{22} + X_{22}^*) \end{vmatrix}, \tag{5.73}
$$

and,

$$\frac{dp_2^*}{dt_2^*} = \frac{1}{|\delta|} \begin{vmatrix} (X_{11} + X_{11}^*) & X_{12} \\ (X_{21} + X_{21}^*) & X_{22} \end{vmatrix}, \tag{5.74}$$

respectively. These expressions have much the same structure as those given above, and also cannot be unambiguously signed. Hence it is not altogether clear that the tariffs on processed wood products that New Zealand faces necessarily result in lower prices for the processed products (nor do they necessarily raise the price of the raw material, as has been suggested).

This analysis has allowed us to place the so-called feedback effect into a true general equilibrium context, without any restrictions on the technology, but does not allow us to make any strong predictions in this respect. As is almost always the case with general equilibrium theory, the answer can only lie in empirical investigation, but this should use general and not partial equilibrium methodology (an approach taken in the following chapters). Notice, however, that we can still say something about the incentives for processing. A home export tax on logs does unambiguously raise the foreign price relative to the home price of logs (driving a wedge between them). Hence, the export tax does raise the attractiveness of processing at home relative to abroad. Similarly, a foreign import tariff on lumber does unambiguously raise the foreign price of lumber relative to the home price, and hence raises the attractiveness of processing in the foreign country relative to the foreign country. Thus, by responding to an escalating tariff with an export tax on the raw material, New Zealand could retrieve some of the processing that the escalating tariff took away. However, we have no way of knowing without performing numerical experiments whether this will be a rational strategy in welfare terms (except in the special case discussed above). Of course, if New Zealand is in fact a small country, then the earlier model results apply, and such a strategy unambiguously lowers welfare in the absence of foreign ownership.

Generalisations of the Model

The underlying model we have developed in the sections above has dimensions that allow us to derive relatively strong and (for the small country at least) unambiguously signed results. However, as is invariably the case in general equilibrium theory, these results do not necessarily continue to hold in higher dimensions, hence the importance of empirical analysis of the issue using a methodology like CGE. To nail this point home, we first consider the effect that an export tax on logs has on factor prices for a small country in a model that is only slightly more complex than the underlying one used above. We will show that it would be incorrect to assume that owners of natural resources are necessarily penalised by the imposition of an export tax on logs, and hence the argument for taxing log exports that relies on the fact that they are produced by foreigners can be shown to rest on fairly shaky foundations. We then utilise the results of Woodland (1982) to fully generalise the underlying model and illustrate some of the problems that may arise.

Allowing for Mobile Capital

We return to the price exogenous case. However, we change the situation under consideration slightly by allowing for capital to be used in all three industries and to be perfectly mobile between the three. Thus, the natural resource-based industry (log production) uses land, labour and capital, the processing industry uses capital and labour in addition to the output of the natural resource-based sector to produce a processed good (lumber), and the third industry (general manufacturing) uses labour and capital only to produce a final good. In our initial equilibrium we maintain our assumption that the country is exporting both logs and lumber in exchange for manufactured goods. Note also that we have maintained our assumption that logs are not utilised by the general manufacturing sector.

Again letting the general manufacturing industry be labelled 0, the natural resource based industry be labelled 1, and the processing industry be labelled 2, we can describe the economy by the following equilibrium conditions:

$$c^0(w,r) = p_0, \tag{5.75}$$

$$c^1(w,r,n) = p_1, \tag{5.76}$$

$$c^2(w,r,p_1) = p_2, \tag{5.77}$$

$$a_{0L}g_0 + a_{1L}g_1 + a_{2L}g_2 = \overline{L}, \tag{5.78}$$

$$a_{0K}g_0 + a_{1K}g_1 + a_{2K}g_2 = \overline{K}, \tag{5.79}$$

$$a_{1N}g_1 = \overline{N}, \tag{5.80}$$

where 5.75-5.77 are the profit maximising conditions and 5.78-5.80 are the factor market equilibrium conditions, as in our more restricted model above. Following the same technique as used above, we are able to obtain the following matrix of equations of change in factor prices:[13]

$$\begin{bmatrix} \theta_{0L} & \theta_{0K} & 0 \\ \theta_{1L} & \theta_{1K} & \theta_{1N} \\ \theta_{2L} & \theta_{2K} & 0 \end{bmatrix} \begin{bmatrix} \hat{w} \\ \hat{r} \\ \hat{n} \end{bmatrix} = \begin{bmatrix} \hat{p}_0 \\ \hat{p}_1 \\ \hat{p}_2 - \theta_{2q_1}\hat{p}_1 \end{bmatrix}, \tag{5.81}$$

which is analogous to equation 5.8 above. We consider only the cases of an export tax and a processing subsidy (the results for other processing incentives follow quite readily). In the case of an export tax, solving for factor prices, we can now determine the impact of an export tax applied to the natural resource based sector (logs) on factor prices as:

$$\hat{w} = \left[-\theta_{2q_1}\theta_{0K}/|\theta|\right]\hat{p}_1, \tag{5.82}$$

$$\hat{r} = \left[\theta_{0L}\theta_{2q_1}/|\theta|\right]\hat{p}_1, \tag{5.83}$$

$$\hat{n} = \left[\{\theta_{0K}(\theta_{1L}\theta_{2q_1} + \theta_{2L}) - \theta_{0L}(\theta_{1K}\theta_{2q_1} + \theta_{2K})\}/\theta_{1N}|\theta|\right]\hat{p}_1, \tag{5.84}$$

where $|\theta| \equiv (\theta_{2L}\theta_{0K} - \theta_{0L}\theta_{2K})$. We therefore know that $|\theta| > 0$ if and only if $\theta_{2L}/\theta_{2K} > \theta_{0L}/\theta_{0K}$. Since we have perfect mobility of the factors between sectors so that wages and rental rates are equalised, the condition holds provided the processing activity is more labour-intensive than production in the manufacturing (import-competing) sector.

We can now conclude the following with respect to the effect of an export tax on real factor rewards from this model. First, from the first two equations we can see that if the processing sector is more labour intensive than the general manufacturing sector, the export tax must cause real wages to rise, and real rental rates on capital to fall. Second, the opposite will hold if the processing sector is more capital intensive than the general manufacturing sector. The export tax causes the returns to owners of capital and labour to move in opposite directions.

The analysis of the impact of the export tax on owners of natural resources can be facilitated by noting that the expressions in the innermost brackets represent the direct and indirect cost shares of labour and capital in the production of one unit of final output of the processed good (lumber), where the cost of labour and capital in the production of the intermediate is taken into account. The numerator of the term inside the brackets will be positive if $(\theta_{1L}\theta_{2q_1} + \theta_{2L})/(\theta_{1K}\theta_{2q_1} + \theta_{2K}) > \theta_{0L}/\theta_{0K}$, therefore the entire expression will be negative if this holds and $|\theta| > 0$. In words, the real return to owners of natural resources will fall with the imposition of an export tax if the processing activity is more labour intensive than production in the import competing sector, and the production of the processed final good (taking into account the cost of labour and capital in the production of the intermediate) is more labour intensive than production in the import competing sector. The real return to owners of natural resources will rise when the factor intensity ranking based on direct and indirect labour requirements is the reverse of the factor intensity ranking based upon the direct capital and labour requirements for the processing activity alone. Clearly, the more labour intensive the production of logs, the more likely it is that the return to natural resource owners will rise if the processing activity is capital intensive. Furthermore, the real return to owners of natural resources will rise by more (or fall by less) the greater the cost share of the intermediate good in the processing activity. This result is interesting from the perspective of more traditional models. The return to natural resources may rise even though it is a specific factor in the industry which sees its price fall, a result that at first seems counter-intuitive. The logic is quite simple, however. The export tax provides an implicit subsidy to processing, which benefits whichever factor is used intensively in processing (say, capital) and harms the other mobile factor (labour). The rise in the return to capital and the fall in the price of logs squeeze the return to natural resources, but the price of labour falls. If log production uses sufficient labour, the return to natural resources may rise.

The impact of a processing subsidy can also be obtained from 5.81. Once again, since equations 5.75 and 5.76 of the system are identical to the case of the export tax, and these two equations uniquely determine the returns to labour and capital, it is clear that both policies have the same effect on the real returns to these factors. Solving for the return to the owners of natural resources yields:

$$\hat{n} = \left[\theta_{2q_1}(\theta_{1L}\theta_{0K} - \theta_{0L}\theta_{1K})/\theta_{1N}|\theta|\right]\hat{p}_1^s, \tag{5.85}$$

The numerator of the term inside brackets will be negative if and only if production of general manufactured goods is more capital intensive than production of logs. The denominator is the same as that from the solution for an export tax given above, and will thus be positive if the processing activity is more labour intensive than general manufacturing. Thus, the condition for the real return to owners of natural resources to rise is that the processing activity and production of logs use the opposite factor intensively when compared with general manufacturing. If both log production and the processing activity are intensive in the same factor the real return to natural resources will fall.

We can compare the effects of the two policies by noting that the expression for the change in the return to owners of natural resources which we obtain with an export tax can be rewritten as:

$$\hat{n} = \left[\{|\theta| + \theta_{2q_1}(\theta_{1L}\theta_{0K} - \theta_{0L}\theta_{1K})\}/\theta_{1N}|\theta|\right]\hat{p}_1. \tag{5.84b}$$

This expression is directly comparable with that for the processing subsidy. If the processing activity is more labour intensive than general manufacturing, then $|\theta|$ is positive. If it is the case that log production is more capital intensive than general manufacturing, then the owners of natural resources would unambiguously gain from a subsidy to domestic processing. It can also be clearly seen that, as in the model above, an export tax of equivalent magnitude would raise the return to owners of natural resources by less, and may lower it (since $|\theta|$ appears as an additive term in the numerator of the term in brackets). Similarly, the owners of natural resources lose by more with an export tax should the returns fall.

These results have some interesting implications for the New Zealand debate surrounding the use of export restrictions on logs for the purpose of increasing domestic processing. The first is that it has commonly been assumed that owners of the forest resource must lose from such a move, and that processors must gain. Hence, in New Zealand we have seen the debate polarised along these lines. However, while the factor used intensively in the processing activity does unambiguously gain from export restrictions on the intermediate, it is not the case that the return to the factor used only in production of the intermediate (here labelled natural resources) must fall. The return may rise or fall depending on factor intensities in it and the other industries.

The impression that the return to owners of natural resources must fall no doubt stems in part from the use of small models such as that utilised above, in addition to those such as that of Lin (1993), which treat the supply of the intermediate good as perfectly inelastic. (Where the supply of the intermediate is exogenously given, the fall in price that the export tax exacts is essentially the same thing as the reduction in real income for the owners of natural resources. This is not the case if we treat the intermediate as a produced good). However, it may in fact be the case that both processors and log producers should have the same interests on this matter, we cannot determine this without detailed knowledge of the industry. While this requires an empirical study such as that which follows, it is interesting to recall the two conditions which influence whether the return to natural resources rises or falls. These are the degree to which log production and processing use the opposite factors intensively relative to manufacturing, and the importance of logs in the processing industry. It is generally considered that the wood processing industry is highly capital intensive, while log production is highly labour intensive (although choices of technology are available in both industries). Moreover, the primary input (roughly 55 percent of total costs) is, not altogether surprisingly, logs. This would suggest that an increase in returns to natural resources is a possibility.

The next interesting issue concerns the argument that an export tax may be used to transfer income from foreign interests (the owners of natural resources in this context) to domestic ones, and thus raise welfare (recall equation 5.37). It should be emphasised again that the real return to owners of natural resources may in fact rise, and an export tax may therefore have an effect which is the opposite of that intended. Somewhat perversely, an export tax or other processing incentive could, for the small economy, result in not only a welfare loss from the reduction in trade, but a further loss through an increase in transfers to foreign owners! Furthermore, if there were significant levels of foreign capital ownership in the rest of the economy (in particular in processing), the situation could be worsened still further, since the return to capital may rise also. Thus the argument in favour of using export taxes to transfer income from foreign owners of the New Zealand forest resource begins to look considerably more tenuous when analysed in a more general framework. It is because of these ambiguities in the direction the results may take that we utilise the method of computable general equilibrium in the following chapters.

A General Specification of Production

In this section, we choose to make use of the generalised model of production with intermediate goods developed by Woodland (1977) and (1982) as the basis for our analysis. There are three reasons for this. The first lies in the obvious benefits of generality in model structure. The second lies in the danger that exists in reading too much into the results of our small scale models, which are indicative but not by any means definitive. The third reason is that our models are in fact a special case of Woodland (1982), as are all of the models discussed at the beginning of this chapter. While few clear results can be obtained from Woodland's approach, we believe discussion of it is of considerable benefit in tying up the model presented here and its relationship to the existing literature, as well as putting the special cases into perspective. It also helps to emphasise the importance of the empirical models that form the second part of this study. The more general approach taken by Woodland (1982) and utilised here thus has considerable appeal.

We again make the standard assumptions that all factors of production are exogenously given and internationally immobile, while all goods (including intermediate goods) are freely tradeable. Once again, using the parlance of Jones and Spencer (1989), the economy we consider can be thought of as consisting of two tiers. In the primary tier of the economy intermediate goods and any final goods which require only the services of primary factors are produced, while processing (in the sense of industries which require intermediate inputs) occupies the secondary tier.

We assume that the production sector consists of *M* industries producing only one product (i.e., we do not allow for joint outputs). We can then characterise production by the following production function:

$$g_j = f^j(x_j, q_j),$$

where g_i is the *M*-dimensional vector of gross output of good *j*, x_j is an *N*-dimensional vector of inputs of primary factors or resources, and q_i is an *M*-dimensional vector of inputs of intermediate goods. It is necessary to distinguish between net and gross outputs of a sector because of the fact that the output of one sector may be used as an input into another. We define net output of *j* as the gross output of good *j* minus the sum of inputs of *j* into all other industries. Net output may be positive (in which case it is available for export or consumption) or negative (in which case it must be imported). Note also that it is quite possible for some goods to be pure final goods which are never used as an input in any industry, and that it is not necessary that production of all goods use intermediate inputs.

The production functions are all assumed to have the following standard properties: they are positive, continuous, and concave for quantities of inputs greater than zero. We further assume that they are linearly homogeneous (i.e., that the production technology exhibits constant returns to scale). If we assume that each industry minimises the cost of production, then we can define the following unit cost function:

$$c^j(w,p) \equiv \min_{x_j,q_j}\{wx_j + pq_j : f^j(x_j,q_j) \geq 1, (x_j,q_j) \geq 0\},$$

where w is an N-dimensional vector of factor prices, and p is an M-dimensional vector of output prices.[14] The unit cost function denotes the minimum cost of all inputs required to produce one unit of gross output, and is positive, linearly homogeneous, and concave in input prices by the properties of the production function. Furthermore, we know by Shephard's lemma that $c_{ij}(w,p) \equiv \partial c^j(w,p)/\partial w_i$ is the optimal input of factor i per unit of gross output of product j. Similarly, we know that $c_{N+k,j}(w,p) \equiv \partial c^j(w,p)/\partial p_k$ is the optimal input of good k per unit of gross output of good j. Note that the optimal usage of primary factors and intermediate goods both depend on both factor returns and intermediate good prices, indicating possible substitutability between factors and intermediates. If an intermediate is never used in industry j, then the derivative of the cost function with respect to its price is zero.

With perfect competition ensuring that each industry makes zero economic profits, we can write the profit maximising conditions for the production sector in terms of the unit cost functions as:

$$c^j(w,p) = p_j \qquad j = 1,\ldots,M. \tag{5.86}$$

The equilibrium conditions are completed by the following factor market condition:

$$\textstyle\sum_{j=1}^{M} c_{ij}(w,p) g_j = v_i \qquad i = 1,\ldots,N, \tag{5.87}$$

where v_i is the endowment of factor i. Given the vector of factor endowments and prices, the equilibrium conditions can be solved for the vector of factor prices and the vector of gross outputs. We may then determine the vector of net outputs as:

$$y_j = g_j - \textstyle\sum_{k=1}^{M} \partial c^k(w,p)/\partial p_j \qquad j = 1,\ldots,M. \tag{5.88}$$

This completes our description of the production specification of the general model. We now turn to examining the characteristics of this specification.

Comparative Statics of Production - Factor Prices

We assume that the economy under consideration is small, and thus prices can be taken as exogenous. With the production side of the economy specified as above, we can answer a number of interesting questions with respect to the impact of export restrictions and processing incentives. We assume for simplicity that the number of factors (N) is equal to the number of goods (M), and that only goods that are produced are used as intermediates. The equilibrium for the production sector is denoted (g^0, w^0). Totally differentiating the equilibrium conditions thus yields:

$$c_w^j(w^0, p)dw = dp_j - c_p^j(w^0, p)dp \qquad j = 1,\ldots,M, \tag{5.89}$$

$$\sum_{j=1}^M c_w^j(w^0, p)dg_j + \left[\sum_{j=1}^M c_{ww}^j(w^0, p)g_j^0\right]dw + \left[\sum_{j=1}^M c_{wp}^j(w^0, p)g_j^0\right]dp = dv \qquad j = 1,\ldots,M, \tag{5.90}$$

where $c_w^j(w^0, p)$ and $c_p^j(w^0, p)$ are the vectors of primary factor and intermediate inputs, respectively, per unit of gross output of good j, and $c_{ww}^j(w^0, p)$ and $c_{wp}^j(w^0, p)$ are matrices of second derivatives of the unit cost function for industry j. Rewriting in percentage change form (e.g., we let $\hat{w} \equiv dw/w$) we have:

$$\theta_w^T \hat{w} = (I - \theta_p^T)\hat{p}, \tag{5.91}$$

$$\lambda\hat{g} + \varepsilon_w \hat{w} + \varepsilon_p \hat{p} = \hat{v}, \tag{5.92}$$

where θ_w is an $N{\times}M$ matrix of factor cost shares, θ_p is an $M{\times}M$ matrix of intermediate cost shares (which are dependent on product and factor prices), I is the identity matrix, λ is an $N{\times}M$ matrix of proportions of total factor use, and ε_w and ε_p are matrices of elasticities of demand for factors with respect to changes in factor prices ($N{\times}N$) and intermediate prices ($N{\times}M$) respectively. A small export tax on an intermediate good will lower the price of that good, with all other prices (and factor endowments) remaining unchanged. Since we have M=N, factor prices are uniquely determined by the profit maximising conditions, and if the tax is sufficiently small that all products previously produced remain in production and that all factors remain fully employed, then the solution for factor prices is given by:

$$\hat{w} = \left[\theta^T\right]^{-1} \hat{p}, \tag{5.93}$$

where $\theta \equiv \theta_w(I - \theta_p)^{-1}$, an $N \times M$ matrix. The matrix θ is thus a matrix of the total of direct and indirect cost of factors necessary to produce one unit of net output.[15] We assume that the matrix θ_w is semi-positive (i.e., that every row and every column contain at least one positive element - implying that each factor is used by at least one industry). Furthermore, it can be shown that the matrix $(I - \theta_p)^{-1}$ exists and is non-negative. Therefore, since $\theta \geq \theta_w$ (this follows immediately from the fact that total factor requirements must be at least as great as direct factor requirements for the production of one unit of a good), θ will also be semi-positive.[16]

We can now partially sign the effects on factor prices of an export tax, since we now know that the fall in price that results from an export tax (say, p_j falls) will cause at least one factor price to fall by a greater percentage, and thus be unambiguously worse off, and at least one factor price to rise, implying that it is unambiguously better off.[17] This can be proven as follows. Since the matrix θ is semi-positive, implying that if $\hat{p}_j < 0$ and all other product prices remain unchanged, at least one factor price must fall to maintain the equality of price and unit cost in industry j. If the first L factor prices fall, then:

$$\sum_{\ell=1}^{L} \theta_{\ell h} \hat{w}_\ell = \hat{p}_h - \sum_{\ell=L+1}^{N} \theta_{\ell h} \hat{w}_\ell \qquad h = 1, \ldots, M, \tag{5.94}$$

where the left hand side is the percentage decrease in the unit costs of industry h caused by the decrease in factor prices, $\hat{p}_h$ is the exogenous percentage decrease in the price of product h(=0 for $h \neq j$) that results from the export tax, and the right hand side summation is the percentage increase in unit cost caused by the increase in the price of factors L+1,...,N. If the production of j uses only one factor, then the left hand side of the equation is zero for all $h \neq j$, and so no other factor prices will rise. If however, production of j uses at least one other factor, then the left hand side of the equation will be negative for some h, and the equation indicates that at least some factor prices, say k, must rise. In either case, the equation indicates that for h=j at least one factor price (say j) must contract by a percentage at least as great as $\hat{p}_j$, since the cost shares cannot exceed unity. Thus, there exist factors j and k such that:

$$\hat{w}_k \geq 0 > \hat{p}_j \geq \hat{w}_j. \tag{5.95}$$

This indicates that at least one factor price will fall by a greater percentage than the fall in price resulting from an export tax, and thus leave the owners of that factor unambiguously worse off, and at least one factor price will rise, as stated above. Exactly which factor prices will rise and which will fall depends of course on the sign structure of the actual matrix under consideration.

Comparative Statics of Production - Outputs

Another issue of obvious interest is the impact of export restrictions on outputs. Intuitively and from the results of our small model above, we would generally expect a lowering of the price of logs to result in a decrease in log production and an increase in processing. However, generalisation of the case with intermediate goods in production raises a number of difficulties in this respect to proving this to be the case. The first is that, in general, it is necessary to consider the prices of inputs that are not produced and must be imported (a situation explicitly excluded from consideration here). The second major difficulty is that there is no unique output concept. In our analysis thus far we have distinguished between gross and net output, but other output measures are also possible (such as value-added). Any or all of these output measures may be of interest for any given problem. Consider first the impact of the a price fall (such as that caused by a log export tax) on gross output. From the comparative static equations above, the solution vector for changes in gross output resulting from an export tax holding factor endowments constant is given by the following when M=N:

$$\hat{g} = -\left[\lambda^{-1}\varepsilon_w(\theta_w^{-1})^T + \lambda^{-1}\varepsilon_p(I-\theta_p^T)^{-1}\right](I-\theta_p^T)\hat{p}, \tag{5.96}$$

where, as above, θ_w is an $N \times M$ matrix of factor cost shares, θ_p is an $M \times M$ matrix of intermediate cost shares (which are dependent on product and factor prices), I is the identity matrix, λ is an $N \times M$ matrix of proportions of total factor use, and ε_w and ε_p are matrices of the elasticities of demand for factors with respect to changes in factor prices ($N \times N$) and intermediate prices ($N \times M$) respectively.[18]

The term $(I-\theta_p^T)\hat{p}$ can be interpreted as the initial or impact percentage change in value-added ignoring any changes that might take place in θ_p as a consequence of changes in factor and intermediate good prices. Unfortunately, the matrix pre-multiplying this term (in square brackets) is of indeterminate sign in general and consequently we cannot sign the changes in gross outputs in response to changes in prices, except under very strict assumptions about the technology. One possibility is to assume that there is no substitution between factors and intermediates (substitution between factors and between intermediates may still be allowed). In this case $\varepsilon_p = 0$ and gross outputs can be shown to be related to a nominal value-added price index in a manner similar to a final good only model.[19] However, as was discussed in the previous chapter, there are reasons for not always assuming fixed proportions technology.

Another possibility is to use the output measure value-added (the excess of revenue in an industry minus the cost of intermediate inputs). It can be shown that where the production technology is separable between factors and intermediates, the relationship between the price of real value-added and the quantity of real value-added is analogous to the relationship between price and output in a final good model. Separability requires that we assume primary factors combine to produce a fictional intermediate product, which is called real value-added, which is then used along with other (non-fictional) intermediate inputs to produce a final output (there is no restriction on the way in which they combine). See Woodland (1982) pp.318-21 for a discussion and illustration of the concepts, and Woodland (1977) pp.527-8 for a formal proof.

Both of these possibilities require very strong assumptions about the underlying technology, which are really a matter for econometric investigation. The issue of whether either assumption is legitimate in the current context will be further examined in Chapter 6, where we present econometric estimates of the elasticities of substitution in the forest products industry. In the absence of any strong restrictions on the technology the change in outputs as a result of price changes is indeterminate. Accordingly, the theoretical details of the matter are not pursued further here, and the impact of processing incentives on gross output is taken to be an empirical matter to be the subject of further examination in the following part of this study.

Summary and Conclusions

This chapter has examined the impact of processing incentives in a static, competitive, general equilibrium framework. It was shown that the relatively clear and unambiguously signed results that can be obtained from an appropriately limited price-exogenous model tend to disappear when the model is extended to include exogenous prices in a multi-country setting, and/or when the dimensions of the model are taken beyond the simple 3×3 case that has concerned us throughout most of this chapter. Many of the questions that we wish to answer simply cannot be handled by a theoretical model (although such an approach may give us significant insight into the nature of the problem). Accordingly, we turn in the following chapter to a computable general equilibrium model.

Notes

[1] The model is of the same scale as that of Suzuki (1978) and Keppler (1985), and Jones and Spencer (1989), although the specification of the model used here is quite different from that used in any of these papers. The main difference is that here we allow for substitution between primary factors and intermediate goods (there are no special assumptions about the technology beyond constant returns to scale). We later consider a model similar to that of Burgess (1976). The models of Burgess (1980) and (1980b) can also be considered as special cases of the model developed by Woodland (1982), but because they include a specific factor in processing, N>M and the factor prices are no longer independent of the factor endowments. While this presents no insurmountable difficulty in deriving the solutions, there are few additional insights to be gained in the current context. We will use the small model of this section to illustrate a number of comparative static results

[2] For a formal proof of the equivalence of these two forms see Dixit and Norman (1980) pp. 44-6.

[3] Note that if we allow for some market power in the log market and maintain the small country assumption for the lumber market (ignoring the possibility of indirect impacts on lumber prices), expression (21) becomes: $E_u du = y_1 dp_1^* + t dy_1$, which describes the decomposition of the welfare effect of an export tax into terms of trade and volume of trade effects. In this case it is possible that welfare may rise with a tax. However, this requires some of the incidence to be pushed onto the world market. The implication is that, while it is possible to increase welfare and processing at the same time, the two objectives are competing (the more successful the policy is in attaining one goal, the less successful it must be in attaining the other), a fact also observed by Keppler (1985).

[4] We are concerned largely with the existence of the foreign ownership and the income transfers that result, rather than the process by which the ownership comes about, i.e., the ownership of the natural resource is exogenous to the

system. The treatment is therefore similar to that used by Hazari and Pattanaik (1980), Bhagwati and Brecher (1980) and Beladi and Marjit (1992).

5 If the country has market power in logs, but none in lumber (and again ignoring the possibility of indirect effects on prices) an export tax would be strictly preferable to either an export subsidy or a processing subsidy. This is because an export or processing subsidy leaves the price paid to domestic log producers equal to the world price, and since exports of logs fall this must rise. Wages still rise as well, but the implication is that the return to owners of natural resources may rise. If this were to occur, a subsidy would have the effect of increasing transfers to foreign owned factors (in addition to the usual transfer to foreign consumers).

6 We assume throughout that the underlying model for the foreign country is identical to that of the home country, with the exceptions that net output of logs is negative, and logs and lumber are imported in exchange for exports of manufactures.

7 Note that the model, at this level of aggregation, is quite similar to that used by Woodland (1983) and Svensson (1984) to deal with symmetric trade in goods and factors. The similarity can be traced back to our assumption that logs are a pure intermediate good, this makes them very similar to a traded factor endowment (in the sense that they do not enter the utility function directly).

8 See Appendix A for a full derivation.

9 Note that this does not necessarily imply that they have to have the same preferences, although this assumption is consistent with the condition. See Woodland (1983) for a discussion of the forms that the preferences may take. Also note that this is in fact a condition for stability of the model, again see Woodland (1983) for details.

10 The proportional change in lumber output can be shown to be given by: $\hat{g}_2 = -(\lambda_{1L}/\lambda_{2L})\hat{N}$.

11 These seemingly paradoxical results are not some theoretical oddity, but have in fact been observed in the real world. Lindsay (1989) examines the case of the Indonesian log export ban, and finds that the large increase in plywood supplied to the world market in fact resulted in a fall in the plywood price (any rise in the price of plywood though increases in log prices was not sufficient to eliminate the decrease in price resulting from increased supply).

12 For details refer to Woodland (1982) pp.278-9.

13 Note that this can be considered as a special case of the model developed by Burgess (1976) where the intermediate good does not enter the production function of the importable. The model can therefore be interpreted as a special case of the three good, three factor model where the services of the third factor enter directly into only one sector. The returns to the two mobile factors are

determined by the last two equations of the system, with the return to natural resources being determined residually.

[14] Note that we maintain the convention of Woodland (1982) in not distinguishing between row and column vectors except where confusion could arise. Thus the term wx_j denotes the inner product of w and x_j.

[15] See Woodland (1982) p.111 for a formal proof.

[16] A formal proof is contained in Lancaster (1968) ch. 6.

[17] In the 2×2 case, the Stolper-Samuelson theorem survives largely intact, with the exception that factor intensities should be defined in terms of total factor input-output coefficients.

[18] See Woodland (1982) p. 322.

[19] See Woodland (1982) pp.322-4 for details.

6 A Computable General Equilibrium Model

Introduction

In the preceding chapter, we have been able to illustrate a number of unambiguously signed theoretical results regarding the impact of various policy options, in the context of a simplified small country model. However, the unambiguous nature of these results tends to break down with the introduction of more goods and factors. The abstract models provide us with the theoretical foundations for the model presented here, and also with a basis for understanding the results of the model as presented in Chapter 7 and those of the GTAP model used in Chapter 8. However, an algebraic model provides us with virtually no information of the likely magnitude of various effects, information that is crucial to the policy debate.

In this chapter we develop and present a small-scale CGE model of the New Zealand economy designed as a quantitative counterpart to the models of the preceding chapter. The main purpose of this chapter is to describe the structure of the model, the data, and the parameter estimation/selection process. Readers less interested in technical aspects of the model may wish to skip large sections of this chapter. We present the results of model simulations in the following chapter.

The chapter is structured into four major sections. In the following section we present an overview of CGE methodology, and a brief review of the existing CGE literature. We then present a detailed description of the small-country model that we utilise here. The final sections consider the process of obtaining parameters, and the process of calibrating the model to a consistent base-year database.

Computable General Equilibrium

While it is beyond the scope of this chapter to provide an exhaustive discussion of the vast and rapidly expanding literature on CGE, in this section we discuss the main features of the methodology, describe some uses to which it has been put, and provide directions to some of the more substantial technical treatments and reviews of the existing literature that are already available.

What exactly is CGE? While the term encompasses an enormous variety of models with widely varying underlying assumptions, analytical structures, implementations, and objectives, all models categorised as CGE are essentially numerical general equilibrium models. They use actual data for the specific economic system under consideration, and solve the system by means of a computer algorithm rather than by formal analytical techniques. As numerical general equilibrium models, CGE models are very closely related to economic theory. They have the great advantage over other multi-sector modelling methods (such as input-output analysis and linear programming models) that they explicitly incorporate basic general equilibrium feedback mechanisms and autonomous decision-making by economic agents.

The first empirical implementation of Walrasian general equilibrium theory is considered to be Johansen (1960). Johansen presented a model of the Norwegian economy, which was solved by linearising the model (in logarithms) about a presumed equilibrium, and solving for proportional changes by matrix inversion. This technique, which is essentially a computerised version of the differential comparative statics approach, had the advantage of minimising data requirements, and required only relatively unsophisticated techniques for solution. Many of the larger and more sophisticated models that followed (such as the Australian ORANI model detailed in Dixon et al. 1982, and the Michigan global trade model), although much larger and more sophisticated, were implemented in much the same way.

The approach of Johansen (1960), while groundbreaking, has several disadvantages. Most importantly, the method requires the assumption that the post-change equilibrium is close to the original. This is because if the contemplated changes in policy are large, or if several of them are expected to be executed simultaneously, a local linearisation may lead to unrealistic conclusions (although it is possible to work around this problem by breaking large changes into multiple smaller ones and solving sequentially). Algorithms that could solve a general equilibrium model by global methods became available in the 1960s. Such algorithms are now available in standardised software packages, such as the GAMS (General Algebraic Modelling System) package created at the World Bank, or the GEMPACK system developed at Monash University as part of the IMPACT project.

There are a number of works available that detail the essential aspects of creating and implementing CGE models. An introduction to the nature of CGE can be found in Chenery and Srinivasan (1989) or Greenaway and Milner (1993), and an introduction to the underlying theory is contained in Devaragan and Lewis (1990) and de Melo (1988). More comprehensive guides are contained in Dervis, de Melo and Robinson (1982) – on whose basic framework the model presented here is based, and de Melo and Tarr (1992).

In terms of the CGE literature, Dervis et al. (1982) classify the existing models into four general categories according to the problems on which they focus. First, there are a number of models of developing and developed countries that focus on issues of international trade, growth, economic structure and/or income distribution. Recent surveys of the models which have been applied to the LDCs can be found in Bandara (1991), Decaluwé and Martens (1988), and Devarajan, Lewis and Robinson (1986). A survey of models applied mainly to developed countries is contained in Shoven and Whalley (1984). These single country models generally have between two and thirty production sectors, with most models having less than ten. They generally consider only one representative consumer (although several models do allow for more than one consumption sector).

In the second category there are models of developed countries that focus on issues of public finance, such as Shoven and Whalley (1974). The third category is the multi-country, international trade models that explore issues concerning the volume and direction of trade and its impact on various regions. Models in this category include the Michigan model, developed by Deardorff and Stern (1990), and the Global Trade Analysis Project (GTAP) model, detailed in Hertel (1997). The final category of Dervis et al. (1982) is models focusing on energy issues.

In such a rapidly growing and diversifying field, there are always a number of new developments that do not fit neatly into such categories. Mercenier and Srinivasan (1994) present a number of models which stretch the boundaries in terms of analysis of macroeconomic stabilisation, imperfectly competitive markets, and inter-temporal trade-offs. Another means of classification is provided by Clarete and Roumasset (1986), who suggest that CGE models should be classified according to their structural framework. They distinguish models that have a clear neo-classical structure, from what they refer to as 'eclectic' models that include non-neo-classical relationships (generally appending a macroeconomic model onto the microeconomic general equilibrium model). They argue that the use of macroeconomic relationships should be avoided in CGE models.

There are a number of valid criticisms of the CGE methodology. Firstly, the models tend to stick very closely to the (generally neo-classical) theory that underlies their structure. This raises the question as to why are CGE models a useful analytical tool in their own right? The main reason is that they are able to provide answers with respect to the effect of policy intervention (or other exogenous shocks) where more traditional techniques such as the comparative statics used in the preceding chapter will not. Algebraic modelling often leads us to an answer that we cannot sign, or easily comprehend in terms of magnitude (and that is often difficult to interpret). A numerical example, based on a carefully chosen and calibrated CGE model, will produce a set of theoretically consistent results that do not suffer from any such ambiguities or difficulties in interpretation. This has obvious advantages in the field of policy analysis, since policy-makers may be more inclined to be influenced by results that they can understand than results that they find incomprehensible.

Another clear advantage a method like CGE has over its theoretical counterparts is the ability to essentially costlessly expand the scale of the model. Most of the major advances in theoretical general equilibrium have come about in the context of small scale models (this is particularly so in the field of international trade theory, where the most influential advances have been made with models using only two goods and two factors). The reason for this is to keep the problem technically tractable. While it is sometimes possible to expand the dimensions of these abstract models, the clarity of the original results is often lost. If a computer is doing the solving, and the results are presented numerically, however, there is no such practical limit on the model dimensions and complexity. The modeller is free to choose a model that is appropriate to the problem at hand, rather than being forced to make arbitrary or unreasonable simplifications in the interests of tractability.

A second related criticism that is not so easily dealt with is that the CGE model is a 'black box', from which answers can mysteriously be drawn. This is related to what is discussed as an advantage of CGE in the preceding paragraph, the ability to solve sophisticated and/or untested theory. This can lead to models where it is unclear exactly what is driving the results, whether the results are sensible, or even whether the model has been implemented without errors in the programming or data. This is a difficult criticism to deal with. Perhaps the only response is that CGE models should always be presented in conjunction with a small-scale abstract model that underlies the structure of the more complex CGE, and that the model-building process should include a thorough testing of the model to determine whether the way it behaves is consistent with what economic theory and our own intuition would lead us to expect. Sensitivity analysis should also be conducted to allow some idea of how the model results vary with changes in key parameters.

CGE models are often regarded as being deterministic. This is true. CGE models are sophisticated theoretical models, and the results are a function of the assumptions made in specifying the model. Perhaps the only response is that we need to ensure that the assumptions being made are appropriate in the context of the issue being considered. CGE is an inductive policy tool. To paraphrase Whalley (1985), the contribution of CGE models is to raise the level of the policy debate, increase the level of understanding of how institutions affect outcomes, challenge the received wisdom, and ultimately offer judgements on issues where there may be competing claims. CGE models are not meant to be perfect descriptions of reality, but to tell a story that is consistent with certain stylised facts, and to provide a consistent framework for the policy debate. The criticism offered above is therefore a criticism of economics as a policy science in general, rather than of CGE models in particular.

A more serious criticism can be made with respect to the degree of rigor in a statistical sense of the CGE approach. With CGE there is no means of providing a statistical check against reality, or even of providing confidence levels for the model result. This lack of statistical credibility is inflamed by the frequent use of 'guesstimate' data for key parameters. Given the extremely large parameter requirements of CGE models, this situation is to a certain extent unavoidable, given that the data necessary to estimate certain economic parameters is not always available. However, in spite of the difficulties involved, there is an increasing tendency to devote more time to estimating these key parameters. We agree that, wherever practical, the key parameters should be estimated rather than guesstimated. However, we believe that the issue is less important than is sometimes made out. It should always be made clear that because CGE models focus on economic fundamentals, they do not predict short term movements well. This is true even if the model contains a large number of econometrically estimated parameters.

Most of the criticisms that have been levelled at the CGE approach are essentially criticisms of the manner in which it has been used, and cautionary notes on the interpretation of results, rather than criticisms of the method itself. None of these criticisms detracts in a serious way from the power of the CGE approach as a tool of policy analysis. Where the economy under consideration is distorted by taxes and rigidities in multiple sectors and significant policy changes are being considered, CGE is the only method that is capable of capturing and quantifying in a sensible way the impact of these measures on all affected economic actors.

A Computable General Equilibrium Model

Overview

The model that we present in this section is a traditional, static, neoclassical CGE trade model of a small, open economy. We agree with the arguments of Clarete and Roumasset (1986) and Johnson (1986), among others, that the neoclassical CGE, with its clear and well understood microeconomic foundations, is the policy tool best suited to facilitating intuitive understanding of economic adjustments and sectoral linkages that result from changes in trade policy. Hence, on the production side, we make the standard assumptions that all markets are perfectly competitive, all firms operate under constant returns to scale, and that all factors of production are fully employed, available in fixed quantities, and internationally immobile. We do not allow for monopoly power in trade. The small country assumption seems the most reasonable starting point for a country the size of New Zealand, in particular since we will use a global trade model in a later chapter for comparison.

We build our model initially as a tool of short run policy analysis, and hence assume that labour is mobile between sectors, while capital and natural resources are not (the implication being that current investment will add to capacity only in future periods).

On the demand side, we assume that households spend a constant proportion of their incomes on each commodity, and that their underlying preferences are identical. Our objective is primarily to analyse the effect of trade policy on sectoral production and overall economic welfare rather than the details of government spending and revenue collection, and hence there is no real role for the government in the model. The government collects tax revenue, but we assume that all of this revenue is transferred back to consumers in a non-distortionary manner. Investment is treated in a similarly simple manner. We make the neoclassical assumption that investment is equal to savings (of the household and government), and hence final demand (the sum of household consumption, government purchases and investment demand) is a fixed proportion of national disposable income (which is equal to GDP minus an initially exogenous trade surplus or net capital outflow).

The most significant difference between this model and the one used in the preceding chapter (apart from the obvious differences in scale and implementation) is that here with respect to trade we incorporate what has become known as the Armington assumption. This means that we treat imports and domestically produced goods in the same industrial category as imperfect substitutes. Fundamentally, the Armington (1969) assumption serves two purposes. The first is that it provides a simple way of allowing for intra-industry trade, a fact which we observe in the empirical data but that cannot be explained within the standard trade framework. The second is that it avoids the possibility of small changes in policy variables leading to an unrealistically large production response. Since we also maintain the small country assumption for both exports and imports, the inclusion of an Armington specification of the tradable sector forces us to reinterpret the small country assumption to mean that the world prices of both exports and imports are exogenous (and hence the terms of trade is fixed).

The Armington assumption has been widely criticised in the literature, and is an assumption that few seem to like using. Perhaps the most important critique is provided by the literature on imperfect competition and trade, in which product differentiation is endogenous, and is associated with the efforts of firms to carve out a market niche for themselves. With Armington, product differentiation is exogenously imposed on the demand side, and its rationale is left largely unexplained. However, despite its limitations, it has achieved widespread acceptance in the CGE literature (to the point where it is incorporated in virtually all CGE models), and is an efficient and practical means of ensuring the model conforms to the empirical evidence. We justify the use of the Armington assumption by virtue of the fact that it stands in for a number of possible explanations for product differentiation that are not sufficiently well understood, and are sufficiently complex, not to warrant direct modelling.

Model Equations

The core equations of our base model are contained in Table 6.1, while Table 6.2 summarises the notation used. A schematic diagram of a simplified version of the model has also been provided (Figure 6.1). We begin with a fairly general specification of production, and discuss the actual industrial aggregation scheme used in the following section. Our economy consists of N industries, each producing a distinct commodity. Industry 1 is forestry, the primary output of which is logs. Industry 2 is wood processing, where logs are transformed into lumber, plywood, etc. Of the remaining industries, $3,\ldots,T$ where $T<N$, are classified as tradable, while the remaining $T+1,\ldots,N$ sectors are non-traded.

Equations 6.1 to 6.3 define the production technology. In the forestry sector (1), production takes the form of a two-level constant elasticity of substitution (CES) function, with capital and labour combining to produce a composite in one level, and this composite combining with

Table 6.1: Equations of the Model

Production Functions

6.1 $$X_1 = \frac{b_1}{\left(1-\sum_{j=1}^{N} a_{1j}\right)} \cdot [\delta_1 \cdot \{\delta_1' \overline{K}_1^{-\rho_1'} + (1-\delta_1') \cdot L_1^{-\rho_1'}\}^{\rho_1/\rho_1'} + (1-\delta_1) \cdot \overline{N}_1^{-\rho}]^{-1/\rho_1}$$

6.2 $$X_2 = \frac{b_2}{\left(1-\sum_{j=2}^{N} a_{2j}\right)} \cdot [\delta_2 \cdot \{\delta_{2K}' \cdot \overline{K}_2^{-\rho_2'} + \delta_{2L}' \cdot L_2^{-\rho_2'} + (1-\delta_{2K}' - \delta_{2L}') \cdot \overline{N}_2^{-\rho_2'}\}^{\rho_2/\rho_2'} + (1-\delta_2) \cdot XQ^{-\rho_2}]^{-1/\rho_2}$$

6.3 $$X_i = \frac{b_i}{\left(1-\sum_{j=1}^{N} a_{ij}\right)} \cdot [\delta_{iK} \cdot \overline{K}_i^{-\rho_i} + \delta_{iL} \cdot L_i^{-\rho_i} + (1-\delta_{iK} - \delta_{iL}) \cdot \overline{N}_i^{-\rho_i}]^{-1/\rho_i} \qquad i = 3,\ldots,N$$

Labour Market Equilibrium

6.4 $$\overline{L} = \sum_{i=1}^{N} L_i$$

6.5 $$W = PN_i \cdot \frac{\partial X_i}{\partial L_i} \qquad i = 1,\ldots,N$$

Processing Demand for Logs

6.6 $$P_1 = PN_2 \cdot \frac{\partial X_2}{\partial XQ_1}$$

Total Intermediate Use

6.7 $$V_1 = \sum_{j=1}^{N} a_{1j} X_j + XQ_1 \qquad j \neq 1$$

6.8 $$V_i = \sum_{j=1}^{N} a_{ij} X_j \qquad i = 1,\ldots,N$$

Utility Function

6.9 $$U = \prod_{i=1}^{N} C_i^{\phi_i}$$

Armington Function

6.10 $$Q_i = \overline{B}_i \cdot [\Delta_i \cdot M_i^{-\mu_i} + (1-\Delta_i) \cdot D_i^{-\mu_i}]^{-1/\mu_i} \qquad i = 1,\ldots,T$$

Demand for Goods

6.11 $$C_i = \phi_i \cdot \frac{NDI}{PCT_i} \qquad i = 1,\ldots,N$$

6.12 $$D_i = DR_i \cdot (C_i + V_i) \qquad i = 1,\ldots,N$$

6.13 $$DR_i = \overline{B}_i^{-1} \cdot \left(\frac{PD_i}{1-\Delta_i}\right)^{-\eta_i} \Bigg/ \left(\Delta_i \cdot \left(\frac{PM_i}{\Delta_i}\right)^{\eta_i \cdot \mu_i} + (1-\Delta_i) \cdot \left(\frac{PD_i}{1-\Delta_i}\right)^{\eta_i \cdot \mu_i}\right)^{\frac{-1}{\mu_i}} \qquad i = 1,\ldots,N$$

6.14 $$M_i = \left(\frac{\Delta_i}{1-\Delta_i}\right)^{\eta_i} \cdot \left(\frac{PD_i}{PM_i}\right)^{\eta_i} \cdot D_i \qquad i = 1,\ldots,T$$

Table 6.1: Equations of the Model (cont.)

Price Equations

6.15	$PM_i = \overline{PW_i} \cdot (1 + tm_i) \cdot XR$	$i = 1,\ldots,T$
6.16	$PD_i = \overline{PWE_i} \cdot XR/(1 + te_i)$	$i = 1,\ldots,T$
6.17	$P_i = \overline{B}_i^{-1} \cdot [\Delta_i^{\eta_i} \cdot PM_i^{(1-\eta_i)} + (1-\Delta_i)^{\eta_i} \cdot PD_i^{(1-\eta_i)}]^{1/(1-\eta_i)}$	$i = 1,\ldots,T$

Net Price Equations

6.18	$PN_1 = PDCT_1 - \sum_{j=1}^{N} a_{j1} \cdot P_j$	$j \neq 2$
6.19	$PN_i = PDCT_i - \sum_{j=1}^{N} a_{ji} \cdot P_j$	$i = 2,\ldots,N$

Income Equations

6.20	$LI = \sum_{i=1}^{N} W \cdot L_i$
6.21	$NI = \sum_{i=1}^{N} \frac{\partial X_i}{\partial \overline{N}_i} \cdot \overline{N}_i$
6.22	$KI = \sum_{i=1}^{N} PN_i \cdot X_i - LI - P_1 \cdot XQ_1 - \sum_{i=1}^{N} P_3 \cdot EN_i$
6.23	$TR = \sum_{i=1}^{T} tm_i \cdot \overline{PW_i} \cdot M_i \cdot XR + \sum_{i=1}^{T} te_i \cdot PD_i \cdot E_i$
6.24	$GDP = LI + KI + TR$
6.25	$NDI = GDP - \overline{F} \cdot XR$

Material Balance Equations

6.26	$X_i = D_i + E_i$	$i = 1,\ldots,T$
6.27	$X_i = D_i$	$i = T+1,\ldots,N$

Balance of Payments

6.28	$\sum_{i=1}^{T} \overline{PW_i} \cdot M_i + \overline{F} = \sum_{i-1}^{T} \overline{PWE_i} \cdot E_i$

Price Normalisation

6.29	$\sum_{i=1}^{N} \Omega_i \cdot P_i = 1$

Table 6.2: Definitions of Variables and Parameters

Industrial Classification:

1	Forestry Industry
2	Wood Processing Industry
3,...,T	Other Traded Goods
T+1,...,N	Non-Traded Goods

Endogenous Variables:

X_i	Gross output
L_i	Sectoral labour employment
XQ_1	Logs used in the processing industry
V_i	Intermediate demand
C_i	Final demand
D_i	Domestic demand for domestic production
DR_i	Domestic use ratio
Q_i	Domestic/import composite
M_i	Imports
U	Social welfare
PM_i	Price of imported goods expressed in domestic currency
PD_i	Price of domestically produced goods, expressed in domestic currency
P_i	Price of the domestic/import composite
XR	Exchange rate
PN_i	Net price expressed in domestic currency
W	Wage rate per unit of labour
LI	Total labour income
NI	Total natural resource income
KI	Total capital income
TR	Tariff and export tax revenue
GDP	Gross domestic product
NDI	Net disposable income
E_i	Exports

Exogenous Variables:

$\overline{PW_i}$	World price of imports
$\overline{PWE_i}$	World price of exports
$\overline{K}_i$	Capital stocks (sector-specific)
$\overline{N}_i$	Stock of natural resources used in industry *i*
$\overline{L}$	Total labour supply
$\overline{F}$	Capital outflow

Table 6.2: Definitions of Variables and Parameters (cont.)

Policy Variables:

tm_i	Ad-valorem import tariffs
te_i	Ad-valorem export taxes

Parameters:

a_{ij}	Intermediate input-output coefficients
b_i	Scale parameter in production of good i
δ_i	Distribution parameters in CES production functions
ρ	Substitution parameters in CES production functions
Δ_i	Distribution parameters in CES Armington functions
μ_i	Substitution parameters in CES Armington functions
η_i	Elasticity of substitution between imports and domestic production
ϕ_i	Consumption expenditure shares
Ω_i	Weights for price index

natural resources to produce the final output. All intermediate goods are assumed to be used in fixed proportions. The two-level CES function is also used in the wood products industry (2), this time with capital, natural resources and labour combining to produce a composite factor at one level, and this composite combining with XQ_1 (the amount of logs used in wood production) in the second level to produce the final output.[1] The production functions of all other goods are assumed to be of the generalized CES form, with capital, natural resources and labour as primary inputs. Once again, all intermediate goods (except logs in the wood processing industry) are used in fixed proportions.

Equations 6.4 and 6.5 define the labour market equilibrium. Equation 6.4 is simply the factor market equilibrium condition, which states that the total supply of labour to the economy is fixed, and that there is no unemployment. Equation 6.5 defines the marginal condition for labour demand in each industry, that the wage rate is equal to the value of the marginal product of labour when producers act to minimise costs.

Equation 6.6 is similar in nature to equation 6.5. Since logs are not used in fixed proportions in the processing industry, it is necessary to specify an industrial demand for logs. We assume that the processing sector pays the prevailing market price for logs, and uses them until the usual marginal condition is satisfied.

Equations 6.7 and 6.8 define intermediate demand for the output of the forestry industry and other sectors respectively.

Turning to the demand structure, equation 6.9 is an aggregate social welfare function of the Cobb-Douglas form.[2] This function determines the overall level of utility for the society, and is the objective function in the model. The demand structure also requires equation 6.10, the Armington function, which aggregates domestic production and imports into a composite good.

Equation 6.11 is final consumption demand, and follows from equation 6.9. Equation 6.12 defines the total domestic demand for domestically produced goods (D_i) as the sum of final demand (C_i) and intermediate demand (V_i), multiplied by the domestic use ratio (DR_i), which is defined as D_i/Q_i, where Q_i is supply of the import/domestic composite. With an Armington aggregation of the CES form (equation 6.10), this is given by equation 6.13.[3]

Also given the Armington aggregation, the demands for imports become derived demands. The problem is to find the ratio of inputs such that the marginal rate of substitution between imports and domestic production equals the ratio of the price of the imported good to the domestically produced commodity. The resulting import demand functions are given by equation 6.14.[4]

Equations 6.15 to 6.19 define gross prices and net prices. Notice that with the Armington specification and intermediate inputs, six prices are required for each good (the world prices of imports and exports respectively, the domestic prices of imports and domestic production, the composite prices, and the net prices). Equation 6.15 relates the domestic prices of imports to the fixed world price (expressed in foreign currency), the difference being an ad-valorem tariff. Similarly, equation 6.16 relates domestic prices to the fixed world prices of exports, with the difference being an ad-valorem export tax. Equation 6.17 requires a little more explanation. Consumption consists of both domestically produced and imported goods. We combine imports and domestic production into a composite good in equation 6.10, which is the Armington function. Since this function takes the CES form in our model, the corresponding composite price (a weighted average of the import and domestic production prices) is also of the CES form (since the CES function is self-dual).[5] Equations 6.18 and 6.19 define net prices, which are equal to the domestic price faced by the producer, less the cost of intermediate inputs. Note that the cost of logs is excluded from the net price calculation for the wood processing sector (equation 6.19), since we allow for some substitution between labour, capital, and logs in this sector.

Equations 6.20 to 6.25 are the income equations, which are largely self-explanatory. Note that in determining the total income accruing to capital we must subtract the value of logs used in processing from the value of net output, in addition to the other factor incomes (since it is not accounted for in the net prices). *GDP* is defined as the sum of all factor incomes and tariff revenue. Net domestic income (*NDI*) is defined as *GDP* less the exogenous capital outflow (the trade surplus). This is the budget constraint faced by the economy.

Equations 6.26 and 6.27 are material balance or market-clearing conditions for the goods markets. The system is closed by the inclusion of the balance of payments function 6.28, the flexible exchange rate being allowed to adjust so as to maintain a fixed capital outflow in the base model (i.e., the current account balance is treated as exogenous). Finally, equation 6.29 defines a price normalisation rule to allow the system to be solved. This is necessary since of the $N+1$ equilibrium conditions, only N are independent. It is necessary to specify a numeraire with which all other relative prices can be compared.[6]

The Long Run

The model we have described above makes the assumption that capital and natural resource stocks are fixed. This is reasonable in the context of short-run simulations. However, in the long run we would expect factors other than labour to be mobile. If for example, we were interested in finding some equilibrium configuration for some future year (without actually modelling the time path to this equilibrium) then it might be sensible to allow the model to determine sectoral capital and natural resource stocks endogenously. Here we run into a theoretical problem.

In the present instance there are three factors, labour, capital and natural resources, but a greater number of goods. In a constant returns to scale competitive model, for arbitrarily given product prices, this implies that only three industries will produce in general. While it is no problem to construct an equilibrium where more goods than factors are produced (we merely need to ensure that all unit value isoquants are tangent to the same factor return plane), such an equilibrium is quite special, and in general a small change in the price vector will cause all but three industries to disappear.[7] Of course, the prices of non-traded goods are not given arbitrarily, but the prices of all traded goods are.

How can we deal with this problem? To a certain extent, the use of the Armington assumption moderates the extent of production changes, since it ensures that there is always domestic demand for domestically produced goods, preventing extreme movements in import-domestic market shares (i.e., it adds more curvature to the production possibility frontier). However, it is still possible that changes in production patterns can be large enough to eliminate the export of many goods, with large quantities of a small number of goods being exported instead.

The obvious 'solution' is to introduce as many factors as there are goods produced. However, in practical terms this is difficult to achieve. It quickly becomes impossible to find ways of separating out a large number of factors from the data available. Given this difficulty, we make the alternative assumption that, in the long run, while labour and capital are mobile, the natural resource component of each industry remains fixed. This is equivalent to assuming that land is the source of long term capital rigidities. Hence, we always have more factors than goods produced, and stability of the model projections is assured.

The assumption that capital can move while natural resources cannot is clearly debatable. The reasonableness of the assumption will depend on the time frame being considered (the longer the time frame, the less likely that any factors will be specific to any particular sector), and the degree of disaggregation of the model (it may be reasonable to assume, for example, that industrial land cannot be used for agriculture even in the medium/long run, but it may be less reasonable to assume that land used for textile production could not be used for other manufacturing purposes). Nevertheless, the assumption is quite reasonable in the context of primary industry (agriculture, mining, forestry, etc.), and manufacturing that has a high natural resource component (extraction industries, for example). It can also be argued that certain industries (e.g., wood processing) require land that is near to the resource that they process due to high transportation costs. Industrial, commercial and residential zoning laws also ensure that some land is restricted in the uses to which it can be put. We therefore feel that for a model that is aggregated to a reasonably high level, and where the time horizon is for the medium-long term, the imposition of some long-term factor movement inflexibility by assuming fixed stocks of natural resources is not unreasonable. We recognise however, that the long-run projections of the model should be interpreted in the light of the assumptions under which they are obtained.[8]

The actual modifications that are necessary to consider the long run impact of policy changes are given in Table 6.3. Essentially this is a matter of adjusting the model closure. We adjust the production functions to include the now mobile capital in equations 6.1 to 6.3, and we add capital market equilibrium conditions 6.30 and 6.31. These are the equivalent of equations 6.4 and 6.5 for the labour market, 6.30 being the full employment condition, and 6.31 being the demand for capital by each industry. This completes our description of the model. Figure 6.1 is a simplified schematic version, with all non-traded goods grouped under the heading 'Non Traded', and all tradables grouped under the heading 'Other Traded'. In the following section we turn to discussing the industrial aggregation and base-year dataset.

Table 6.3: Adjustments for the Long Run

Production Functions:

6.1 $$X_1 = \frac{b_1}{\left(1-\sum_{j=1}^{N} a_{1j}\right)} \cdot [\delta_1 \cdot \{\delta_1' K_1^{-\rho_1'} + (1-\delta_1') \cdot L_1^{-\rho_1'}\}^{\rho_1/\rho_1'} + (1-\delta_1) \cdot \overline{N}_1^{-\rho}]^{-1/\rho_1}$$

6.2 $$X_2 = \frac{b_2}{\left(1-\sum_{j=2}^{N} a_{2j}\right)} \cdot [\delta_2 \cdot \{\delta_{2K}' \cdot K_2^{-\rho_2'} + \delta_{2L}' \cdot L^{-\rho_2'} + (1-\delta_{2K}' - \delta_{2L}') \cdot \overline{N}_2^{-\rho_2'}\}^{\rho_2/\rho_2'} + (1-\delta_2) \cdot XQ^{\rho_2}]^{-1/\rho_2}$$

6.3 $$X_i = \frac{b_i}{\left(1-\sum_{j=1}^{N} a_{ij}\right)} \cdot [\delta_{iK} \cdot K_i^{-\rho_i} + \delta_{iL} \cdot L_i^{-\rho_i} + (1-\delta_{iK} - \delta_{iL}) \cdot L_i^{-\rho_i}]^{-1/\rho_i} \qquad i = 3,\ldots,N$$

Capital Market Equilibrium:

6.30 $$\overline{K} = \sum_{i=1}^{N} K_i$$

6.31 $$R = PN_i \cdot \frac{\partial X_i}{\partial K_i} \qquad i = 1,\ldots,N$$

Endogenous Variables:

K_i Capital employed in industry *I*

R Rental rate on capital

Exogenous Variables:

$\overline{K}$ Total capital stock

Industrial Aggregation and Base Year Data

The primary source of data for a CGE model is an input-output table. In New Zealand, the latest year for which an input-output table has been constructed is 1993. This data is only available at the 25 Sector SNA (System of National Accounts) level. More detailed data is somewhat older, dating back to 1987. As the New Zealand economy underwent significant reform between these periods, we feel that it is more appropriate to use the 1993 data.

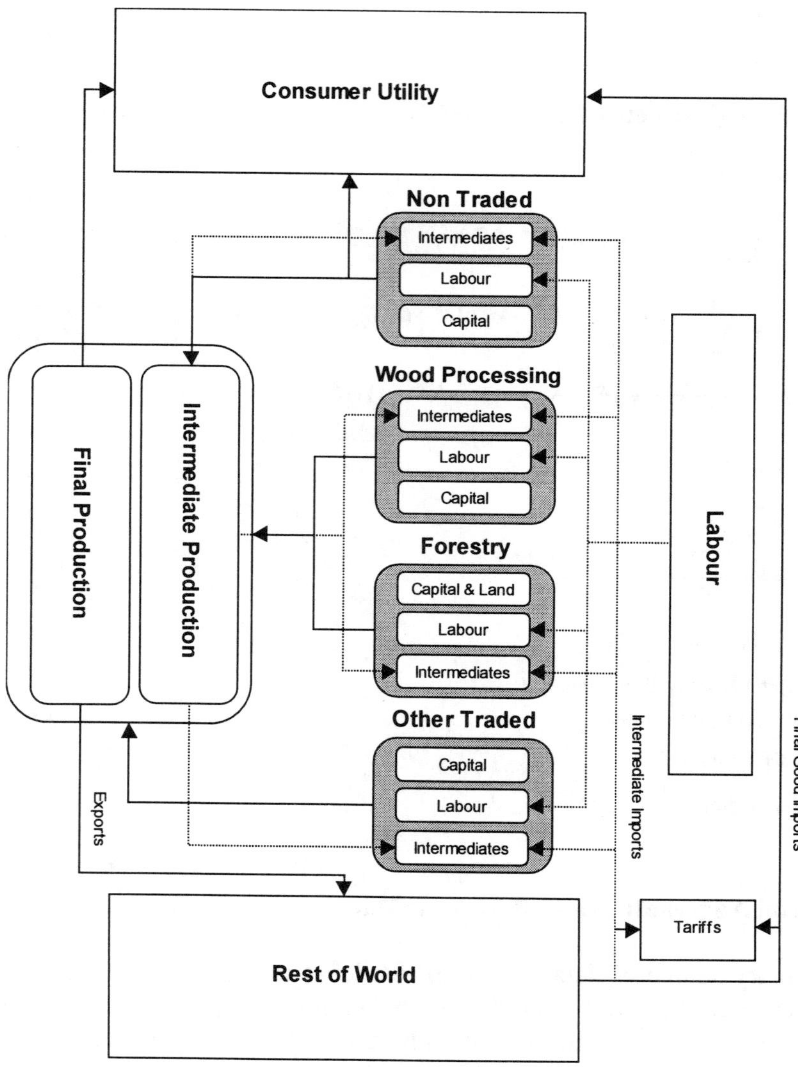

Figure 6.1: Schematic Representation of the Model

The next major issue is the selection of a level of aggregation for the current model. The issue is influenced by the data available, and the use to which the model is to be put. Technical issues are relatively unimportant. We choose to follow a trend in the CGE literature of aggregating the model into a relatively small number of sectors to enable us to focus more clearly on those of particular interest (i.e., the forestry and wood processing industries).

Table 6.4: Industrial Aggregation

Industry Group	Industries Incorporated	SNA
Traded Goods		
Agriculture, Fishing & Hunting	Agriculture	1
	Fishing and Hunting	2
Forestry and Logging	Forestry and Logging	3
Mining and Quarrying	Mining and Quarrying	4
Light Manufacturing	Food, Beverages and Tobacco	5
	Textile and Apparel Manufacturing	6
Wood Product Manufacturing	Wood and Wood Product Manufacturing	7
Heavy Manufacturing	Paper, Printing and Publishing	8
	Chemicals, Petrol, Rubber, etc.	9
	Non-metallic Mineral Product Manufacturing	10
	Basic Metal Product Manufacturing	11
	Fabricated Metal Product Manufacturing	12
	Other Manufacturing	13
Traded Services	Trade, Restaurants and Hotels	16
	Transport and Storage	17
	Communication Services	18
	Finance, Insurance, etc.	19
Non Traded Goods		
Non Traded Services	Electricity, Gas, and Water	14
	Ownership of Owner-occupied Dwellings	20
	Community, Social Services, etc.	21
	Central and Local Government Services	22, 23
	Private Non-profit Services to Households	24
	Domestic Services of Households	25
Construction	Construction	15

Source: New Zealand System of National Accounts

Table 6.5: Input-Output Table 1993 ($NZ1993 millions)

	(1)	(2)	(3)	(4)	(5)	(6)
(1) Agriculture, Fishing and Hunting	1976	19	0	6528	5	29
(2) Forestry and Logging	5	608	0	4	314	152
(3) Mining and Quarrying	8	9	275	51	1	1753
(4) Light Manufacturing	200	10	1	3552	81	147
(5) Wood Product Manufacturing	24	2	1	22	368	169
(6) Heavy Manufacturing	1344	104	96	1644	310	8550
(7) Traded Services	1730	220	265	2943	459	3853
(8) Non Traded Services	423	11	23	423	48	664
(9) Construction	386	8	45	123	41	220
(10) Sub Total (1)-(9)	6096	991	706	15290	1627	15537
(11) Labour	1140	121	202	2700	554	4131
(12) Capital	567	337	292	1198	197	1924
(13) Natural Resources	2876	587	515	141	108	1112
(14) Sub Total (11)-(13)	4583	1045	1009	4039	859	7167
(15) Grand Total (14)+(10)	10679	2036	1715	19329	2486	22704

Source: Statistics New Zealand (1995)

The actual aggregation scheme which we utilise is presented in Table 6.4. We map the 25 Sector SNA to nine industrial groups, three in each of the primary, secondary, and services sectors. Primary production is separated out into agriculture and fishing, forestry and logging, and mining and quarrying. Secondary or manufacturing production is separated into light manufacturing, wood product manufacturing (the focus), and heavy manufacturing. Finally, the services sector is separated into three groups, traded services, non-traded services, and construction. The selection criteria for whether a service was counted as traded or non-traded was less than five percent of total supply being either exports or imports. Note that there is no attempt to provide any special explanation for trade in services in this model, tradable services are treated in exactly the same way as any other tradable good. The construction industry is separated out from the other non-traded goods because it is the primary recipient of processed wood products, and thus the most affected by changes.

Table 6.5 presents the modified input-output table for New Zealand 1993 that we use to calibrate our model. The table has been constructed from the 25 SNA Sector Inter-Industry Transactions 1993, and the Imports

Table 6.5: Input-Output Table (1993) (cont.)

(7)	(8)	(9)	(10) Sub Total (1)-(9)	(11) C	(12) I	(13) E	(14) M	(15) TR	(16) Grand Total
233	81	2	8873	544	62	1568	335	33	10679
16	9	0	1108	32	410	492	6	0	2036
42	286	16	2441	12	44	293	1074	1	1715
1536	399	56	5982	5597	538	9654	2370	72	19329
65	115	609	1375	395	187	672	134	9	2486
4180	2710	2758	21696	4532	5250	4257	12901	130	22704
14897	4112	1698	30177	16506	2194	6171	4879	251	49918
1412	3526	288	6818	27193	80	0	0	0	34091
738	1507	1852	4920	111	4465	0	0	0	9496
23119	12745	7279	83390	54922	13230	23107	21699	496	152454
13756	11610	1552	35766						
11670	2385	600	19170						
1373	7351	65	14128						
26799	21346	2217	69064						
49918	34091	9496	152454						

into Industry Transactions of the same year, according to the aggregation scheme described above. The left hand section of the table summarises the inter-industry flow of goods, with each entry reflecting the flow of goods or services from the row to the column sector, and the flow of funds or payments from the column to the row sector. The right hand side summarises the allocation of final demand among the commodities. The sum of industry columns (total supply, row 15) is equal to total demand, the sum of intermediate (column 10) and final demand (the sum of columns 11-13, less the sum of columns 14 and 15) in column 16, by definition. *GDP* is can be obtained from the sum of factor returns (entry 14,10) plus tariff revenue, or from expenditure as ($C+I+X-M$).

Row 13, which gives the contribution of natural resources to each industry probably requires more explanation. Since operating surpluses (capital returns) are the balancing item in the input-output table, and the returns to natural resources are not observed, we need some method of separating out the incomes of natural resource owners and owners of capital. For the agricultural sector, we accomplish this by using rural land area and price data (available from Valuation New Zealand, 1996). Using these figures to estimate the value of agricultural land in 1993, we estimate the return by multiplying by the average interest rate in 1993 (i.e., we interpret the opportunity cost of holding agricultural land as the interest which could have been received has the land been sold for its market value). The return to capital follows residually.

Unfortunately, we cannot use the same technique for other industries since land use by each industry is unknown. Perhaps more importantly, we need to consider the value of the natural resources used in production rather than merely the land. We choose to estimate the proportion of income accruing to natural resources using data on the asset holdings of listed companies. We take those assets which can be classified as natural resources which are held by representative firms on the listed exchange for each category at the 25 sector level, and from these estimate the aggregated figures presented in Table 6.5. Notice that for some sectors the level is considerably higher than others. The level for mining and quarrying is particularly high, reflecting the value of the minerals being extracted. Heavy manufactures is also relatively high, largely reflecting the inclusion of the oil and gas extraction industries, while tradable services, construction and light manufacturing have relatively low contributions from natural resources. This method of separating operating surpluses into a capital and natural resource component is far from perfect, but is the best we have available. It is better than the method used in the GTAP3 database, where the residual is split 50-50 in the case of New Zealand.

Parameter Selection

Elasticity of Substitution in Wood Processing

Since the elasticity of substitution in the wood processing industry plays a key role in the model, we devote this section to its econometric estimation.

Where there are only two factors of production (capital, K, and labour, L), the elasticity of input substitution under constant output measures the curvature of an isoquant. However, there is no standard definition of the elasticity of input substitution when the number of factors of production is greater than two. McFadden (1978) lists the most common definitions as the direct partial elasticity of substitution (DES), the shadow elasticity of substitution (SES), and the Allen (Uzawa) partial elasticity of substitution (AES). The DES is a direct application of the two factor formula to each pair of inputs. It can thus be interpreted as a short-run elasticity in which the quantities of other factors are held constant. The SES is the elasticity calculated by applying the formula to each pair of factors, fixing the average cost and prices of other factors. It can be viewed as a long-run elasticity with factors freely traded at fixed prices. The AES is calculated where prices of output and factors other than the relevant ones are held constant.

For our purposes, the AES has two advantages over the other possible definitions. Firstly, it can be directly linked to the price elasticity of factor demand. Secondly, it can be related directly to the elasticities of substitution in a two-level CES function when inputs belong to different groups. Because of these desirable properties, we estimate the AES here.[9]

We follow the estimation procedure used by Berndt and Christensen (1973) and Berndt and Wood (1975). We assume that there exists in the New Zealand forest products sector a twice differentiable aggregate production function relating the flow of output (*Y*) to the services of three inputs: capital (*K*), labour (*L*), and raw materials (*W*).[10] Corresponding to such a production function there exists a cost function that reflects the production technology.

For the purpose of estimation we must employ a specific functional form for the cost function. Shephard duality assures us that if a cost function satisfies linear homogeneity and concavity in factor prices, there exists a unique production function. Therefore, we can estimate the cost function and the factor demand functions without knowing the specific functional form of the production function. The functional form must be capable of approximating the true function to a desired order of accuracy, and must result in estimating forms that are linear in parameters. We specify a general form that places no *a priori* restrictions on the Allen partial elasticities of substitution, and that can be interpreted as a second order approximation to an arbitrary twice differentiable cost function. Following Berndt and Wood (1975), we choose to use the translog (transcendental logarithmic) cost function introduced by Christensen, Jorgensen and Lau (1971).

Write the translog cost function as:

$$\begin{aligned}\ln C = {} & \alpha_0 + \alpha_K \ln P_K + \alpha_L \ln P_L + \alpha_W \ln P_W \\ & + 1/2\,\gamma_{KK} (\ln P_K)^2 + 1/2\,\gamma_{LL} (\ln P_L)^2 + 1/2\,\gamma_{WW} (\ln P_W)^2 \\ & + 1/2\,\gamma_{KL} \ln P_K \ln P_L + 1/2\,\gamma_{LK} \ln P_L \ln P_K \\ & + 1/2\,\gamma_{LW} \ln P_L \ln P_W + 1/2\,\gamma_{WL} \ln P_W \ln P_L \\ & + 1/2\,\gamma_{KW} \ln P_K \ln P_W + 1/2\,\gamma_{WK} \ln P_W \ln P_K \\ & + \beta_K \ln P_K \ln Y + \beta_L \ln P_L \ln Y + \beta_W \ln P_W \ln Y \\ & + \delta_1 \ln Y + 1/2\,\delta_2 (\ln Y)^2,\end{aligned} \tag{6.32}$$

where Y is total output of the forest products industry, the P_i are the unit costs of the inputs, and C is total cost.

We can simplify the function by imposing symmetry and constant returns to scale conditions. These imply that:

$$\gamma_{ij} = \gamma_{ji} \qquad i, j = K, L, W, \tag{6.33}$$

$$\begin{aligned} \delta_1 &= 1, \\ \delta_2 &= 0, \\ \beta_K &= \beta_L = \beta_W = 0. \end{aligned} \tag{6.34}$$

This allows us to rewrite equation 6.32 in its restricted form as:

$$\begin{aligned} \ln C = {} & \alpha_0 + \alpha_K \ln P_K + \alpha_L \ln P_L + \alpha_W \ln P_W \\ & + 1/2\, \gamma_{KK} (\ln P_K)^2 + 1/2\, \gamma_{LL} (\ln P_L)^2 + 1/2\, \gamma_{WW} (\ln P_W)^2 \\ & + \gamma_{KL} \ln P_K \ln P_L + \gamma_{LW} \ln P_L \ln P_W + \gamma_{KW} \ln P_K \ln P_W \\ & + \ln Y. \end{aligned} \tag{6.35}$$

Further, linear homogeneity in prices imposes the following restrictions on equation 6.35:

$$\begin{aligned} & \alpha_K + \alpha_L + \alpha_W = 1, \\ & \gamma_{KK} + \gamma_{KL} + \gamma_{KW} = 0, \\ & \gamma_{KL} + \gamma_{LL} + \gamma_{LW} = 0, \\ & \gamma_{KW} + \gamma_{LW} + \gamma_{WW} = 0. \end{aligned} \tag{6.36}$$

Assuming perfect competition in the factor markets, we can treat input prices as fixed. Given the level of output, cost minimising input demand functions are derived by logarithmically differentiating equation 6.35 to obtain:

$$\frac{\partial lnC}{\partial lnP_i} = \frac{\partial C}{\partial P_i}\frac{P_i}{C} = \alpha_i + \sum_i \gamma_{ij}\, lnP_j \qquad i, j = K, L, W,$$

and then utilising Shephard's Lemma:

$$x_i = \frac{\partial C}{\partial P_i} \qquad i = K, L, W,$$

to obtain the following equations:

$$\theta_K = \frac{P_K K}{C} = \alpha_K + \gamma_{KK} \ln P_K + \gamma_{KL} \ln P_L + \gamma_{KW} \ln P_W,$$
$$\theta_L = \frac{P_L L}{C} = \alpha_L + \gamma_{KL} \ln P_K + \gamma_{LL} \ln P_L + \gamma_{LW} \ln P_W, \tag{6.37}$$
$$\theta_W = \frac{P_W W}{C} = \alpha_W + \gamma_{KW} \ln P_K + \gamma_{LW} \ln P_L + \gamma_{WW} \ln P_W,$$

where the θ_t are the cost shares of the inputs in the total cost of producing *Y*.

Uzawa (1962) has derived the Allen partial elasticities of substitution (AES) between inputs *i* and *j* as:

$$\sigma_{ij} = \frac{CC_{ij}}{C_i C_j},$$

where:

$$C_i = \frac{\partial C}{\partial P_i},$$

and,

$$C_{ij} = \frac{\partial^2 C}{\partial P_i \partial P_j}.$$

With the translog cost function the AES are:

$$\sigma_{ii} = \frac{\gamma_{ii} + \theta_i^2 - \theta_i}{\theta_i^2} \qquad i = K, L, W, \tag{6.38}$$

and,

$$\sigma_{ij} = \frac{\gamma_{ij} + \theta_i \theta_j}{\theta_i \theta_j} \qquad i, j = K, L, W, \quad i \neq j. \tag{6.39}$$

Note that the AES are not constrained to be constant, but may vary with the values of the cost shares.

The price elasticity of demand for factors has been shown by Allen (1938) to be analytically related to the AES in the following way:

$$E_{ij} = \theta_j \sigma_{ij} . \qquad (6.40)$$

We proceed by estimating the parameters of equation system 6.37, without imposing constant returns to scale and symmetry conditions, to test the appropriateness of the restrictions. We then re-estimate the parameters with the constant returns to scale and symmetry conditions imposed. The parameters of the system can be estimated equation by equation using OLS, and this will produce consistent and unbiased estimates. However, the procedure will not be efficient due to the correlation across error terms arising from the fact that the cost shares sum to one. Given the relatively short series of data available (48 observations), a method which provides a gain in efficiency over OLS is preferable. A further problem is that the cross equation symmetry conditions cannot be imposed and tested using equation by equation OLS. An alternative and more efficient procedure is to apply the method of seemingly unrelated regressions (SUR) to any two of the three factor demand equations (since of the twelve estimated parameters, only eight are free; the parameter estimates from any of the two equations can be derived from the estimates of the other two equations). This procedure is also known as Zellner-efficient estimation. The problem with this approach is that the estimates of the translog parameters are not independent of the choice of which equations to estimate. To eliminate the problem of equation choice dependency, we choose to iterate the ZEF procedure (IZEF). Kmenta and Gilbert (1968) have shown that if one iterates the ZEF procedure, the parameter estimates will converge to the maximum likelihood estimates (which Barten, 1969, has shown to be independent of the equation omitted). One problem with this method of estimation is that, while it may be reasonable to assume that the supply of inputs is perfectly inelastic at the level of the individual firm, and that therefore prices may be taken as fixed, at the more aggregated industry level input prices are less likely to be exogenous. Ideally, this possible simultaneity should be taken into account in the estimation procedure. Berndt and Wood (1975) deal with the problem by utilising an iterated three-stage least squares procedure. This is not pursued here.

The data used in the estimation is set out in Appendix C. All data is quarterly, for years 1984 to 1995, and is derived from the Economic Survey of Manufacturing for NZSIC division 33 (which corresponds exactly to the classification of Wood Product Manufacturing used in the input-output table). We define net output as the value of sales net of the costs of all intermediates (which are assumed to be used in fixed proportions in our model) except logs. The labour cost shares are obtained from the total salaries and wages statistics of the survey. The residual from purchases after other intermediates have been netted out is interpreted as the cost of log inputs. The residual from net output after subtracting labour and log costs is interpreted as the cost of capital.

The price of labour is the average hourly wage rate, obtained by dividing the total salaries and wage statistics of the survey by total hours worked. This can thus be interpreted as the cost of labour providing that the composition of the workforce has not changed over the survey period. The price of logs is obtained from the quarterly log export price index (since we are not interested in the intercept term of the regression, and the other terms are unaffected by any scaling of the regressors, the use of a price index is valid). The price of capital is proxied by the interest rate on 90 day bank bills from 1987, and the average return on 90 day commercial bills prior to this date.

We begin by estimating the system without imposing constant returns to scale and symmetry conditions, and test the following restrictions:

$$\begin{aligned} &\gamma_{LW}^{L} = \gamma_{LW}^{W}, \\ &\gamma_{KW}^{K} = \gamma_{KW}^{W}, \\ &\gamma_{KL}^{K} = \gamma_{KL}^{L}, \end{aligned} \tag{6.41}$$

where a superscript refers to the demand function from which the parameter is estimated. We arbitrarily choose to estimate the system using the *L* and *W* equations from system 6.37. Since we only estimate two equations, and not all of the parameters in the restrictions 6.41 are in contained in the estimating equations, we use equivalent conditions:[11]

$$\begin{aligned} &\gamma_{LW}^{L} - \gamma_{LW}^{W} = 0, \\ &\gamma_{KL}^{L} = -(\gamma_{LL}^{L} + \gamma_{LW}^{L}), \\ &\gamma_{KW}^{W} = -(\gamma_{LW}^{W} + \gamma_{WW}^{W}). \end{aligned} \tag{6.42}$$

Table 6.6 below displays the results of the estimation procedure, using the labour and raw materials equations from 6.37 as described above. Note that there are two estimates of all cross parameters, since the symmetry conditions have not been imposed. The R-square statistic for the labour demand function is 0.6911, and for the raw materials input demand function is 0.2345. The F-statistic for restrictions 6.42 is 1.789, thus we cannot reject the null hypothesis that restrictions 6.42 are true. For the remainder of our estimation we maintain the constant returns to scale and symmetry conditions.

Table 6.6: Estimation Results (Processing)

Parameter	Not Restricted	(t-ratios)		Restricted	(t-ratios)	
α_K	-0.1831	-0.295		0.4385	6.325	*
α_L	1.0573	6.424	*	0.7148	9.877	*
α_W	0.1241	0.225		-0.1534	-1.766	***
γ_{KK}^{K}	0.0500	1.067		0.0051	0.304	
γ_{KL}^{K}	0.0557	0.629		0.0355	6.268	*
γ_{KW}^{K}	0.0245	0.309		-0.0406	-2.709	*
γ_{KL}^{L}	0.0094	0.752		0.0355	6.268	*
γ_{LL}^{L}	0.0102	0.433		0.0351	1.897	***
γ_{LW}^{L}	-0.1014	-4.780	*	-0.0706	-4.467	*
γ_{KW}^{W}	-0.0595	-1.415		-0.0406	-2.709	*
γ_{LW}^{W}	-0.0661	-0.833		-0.0706	-4.467	*
γ_{WW}^{W}	0.0768	1.081		0.1112	5.891	*

*: Significant at the 1% level
**: Significant at the 5% level
***: Significant at the 10% level

The results of re-estimating equation system 6.37 with constant returns to scale and symmetry imposed are also shown in Table 6.6. The R-square statistic for the labour demand function is 0.6543, and for the raw materials input demand function is 0.2299. The Durbin-Watson d statistics for the equations are 1.3516, 1.9257 and 1.8047 for the labour, raw materials, and capital equations respectively. Therefore, for all equations we cannot reject the null hypothesis of no autocorrelation.

Because the translog function does not satisfy concavity and positivity restrictions globally, it is also necessary to check whether the fitted translog function is concave in input prices and if its input demand functions are strictly positive at each observation. Positivity is satisfied if the fitted cost shares are positive. We check the fitted cost shares based on the IZEF parameter estimates and find that the positivity conditions are satisfied at each observation. Concavity of the cost function is satisfied if the Hessian matrix, based on the IZEF parameter estimates, is negative semi-definite. We find that the concavity condition is also satisfied at each observation. We conclude that the cost function is well-behaved over the region 1984-1995.

Using these fitted values, and the cost shares in Appendix C, it is possible to calculate the Allen elasticities of substitution using equations 6.38 and 6.39. The annual averages of these estimates are given in Table 6.7. All of the own group elasticities are negative, as expected. The estimates of the elasticity of substitution between capital and raw materials are all positive, indicating that capital and logs are substitutes in production. The estimates of the elasticity of substitution between capital and labour, and between raw materials and labour are also all positive. The estimated elasticities between capital and logs, and between labour and logs are quite stable at around 0.5, while the estimates of the elasticity of substitution between labour and capital appears to be quite stable at around 1.4.

Table 6.7: Estimated AES (Processing)

Year	σ_{LL}	σ_{KK}	σ_{WW}	σ_{LK}	σ_{LW}	σ_{KW}
1984	-1.157	-2.625	-1.093	1.309	0.443	0.501
1985	-1.237	-2.998	-0.974	1.355	0.492	0.516
1986	-1.117	-5.609	-0.861	1.551	0.560	0.300
1987	-1.178	-2.952	-1.031	1.340	0.463	0.490
1988	-1.249	-2.931	-1.019	1.349	0.388	0.484
1989	-1.308	-1.972	-1.111	1.272	0.247	0.514
1990	-1.298	-2.656	-1.024	1.332	0.430	0.533
1991	-1.247	-4.676	-0.811	1.508	0.556	0.420
1992	-1.387	-4.053	-0.765	1.486	0.549	0.500
1993	-1.597	-3.313	-0.796	1.455	0.489	0.570
1994	-1.630	-2.945	-0.819	1.420	0.465	0.592
1995	-1.645	-2.824	-0.796	1.410	0.484	0.608

Table 6.8 gives the estimates of price elasticity of input demand, calculated using formula 6.40. The results indicate that both capital and labour are relatively responsive to changes in their own prices, the own price elasticities being approximately -0.5 and –0.7, respectively. Raw material

is somewhat more inelastic, at around –0.3. The cross-price elasticities merely confirm what we already know from the input substitution elasticities discussed above.

Table 6.8: Estimated Price Elasticities of Demand (Processing)

Year	E_{LL}	E_{KK}	E_{WW}	E_{KL}	E_{WL}	E_{WK}	E_{LK}	E_{LW}	E_{KW}
1984	-0.492	-0.707	-0.326	0.557	0.189	0.138	0.357	0.135	0.150
1985	-0.505	-0.731	-0.331	0.555	0.202	0.129	0.336	0.169	0.176
1986	-0.483	-0.784	-0.320	0.682	0.247	0.073	0.263	0.220	0.102
1987	-0.495	-0.721	-0.323	0.565	0.195	0.128	0.343	0.153	0.156
1988	-0.507	-0.698	-0.299	0.550	0.160	0.140	0.370	0.138	0.148
1989	-0.516	-0.635	-0.275	0.503	0.097	0.178	0.439	0.077	0.133
1990	-0.514	-0.701	-0.323	0.531	0.173	0.151	0.369	0.144	0.171
1991	-0.507	-0.777	-0.316	0.616	0.227	0.089	0.280	0.227	0.161
1992	-0.528	-0.771	-0.313	0.565	0.209	0.104	0.296	0.232	0.206
1993	-0.552	-0.731	-0.314	0.504	0.170	0.145	0.349	0.203	0.228
1994	-0.556	-0.718	-0.315	0.484	0.159	0.156	0.365	0.191	0.234
1995	-0.558	-0.723	-0.320	0.478	0.164	0.156	0.362	0.196	0.245

Finally, we follow the procedures set out in Berndt and Christensen (1973) in testing the functional separability of the estimated parameters. This can be interpreted as testing whether the functional form which we have specified (the two-level CES) is appropriate. Global separability is defined as the case where $\sigma_{KL} = \sigma_{LW} = \sigma_{KW} = 1$ (i.e., where the function is Cobb-Douglas in form). By equation 6.38 this is equivalent to testing the joint hypothesis that $\gamma_{KL} = \gamma_{LW} = \gamma_{KW} = 0$. The F-statistic for this test is 42.48, and we can therefore decisively reject the null hypothesis of global separability. We now test the linear separability hypotheses that $\sigma_{KL} = \sigma_{LW} = 1$, $\sigma_{LW} = \sigma_{KW} = 1$, and $\sigma_{KL} = \sigma_{KW} = 1$. The resulting F-statistics are 43.82, 26.58, and 17.77 respectively, and we therefore reject all three linear separability hypotheses. The estimated function is clearly not of the Cobb-Douglas form. Also of interest is the non-linear hypothesis that $\sigma_{LW} = \sigma_{KW} \neq 1$. This is important since, by using the two-level CES, we are assuming that labour and capital combine to form a composite (value-added) which then combines with logs. For this to be valid, the above hypothesis should be true. We test to see whether the joint hypothesis that $\gamma_{WW} = \gamma_{KW}^2 / \gamma_{KK}$ and $\alpha_W = \alpha_K \gamma_{KW} / \gamma_{KK}$

holds (see Berndt and Christiansen, 1973, for derivation of this testable hypothesis). The χ^2 statistic for this hypothesis is 0.47, and hence we cannot reject the null hypothesis that the elasticities are equal. We can however reject the hypothesis that $\sigma_{LK} = \sigma_{LW}$ (the χ^2 statistic for this hypothesis is 7.04). Given that this form of non-linear separability exists, no other form can. We therefore terminate our testing here. The results have given us some confidence that the functional form we are using in the model is appropriate.

The final necessary step is to relate the AES to the elasticities of substitution used in the two-level CES function. The outer elasticity (σ) provides no great difficulty. This is simply the elasticity of substitution between value-added (the labour/capital composite), and logs. Since we have shown that $\sigma_{LW} = \sigma_{KW}$, we specify this elasticity as the average of these two estimates. For our base model we use the average value of 0.483.

The inner elasticity (σ') is a little more difficult to deal with. Sato (1967) has shown that the AES and the inner elasticity of substitution in a two level CES function are related in the following way:

$$\sigma_{LK} = \sigma + \frac{1}{\theta_L + \theta_K}(\sigma' - \sigma).$$

The calculated σ' appear to be quite stable. For our base model, we use the average value of 0.839 (we perform some sensitivity analysis in the following chapter). This completes our discussion of the parameter estimation for the wood products industry of the model.

Elasticity of Substitution in Forestry and Logging

The other key elasticity of interest in this study is the elasticity of substitution between primary factors in the forestry and logging industry. Like the wood processing industry discussed in some detail in the sub-sections above, we have utilised a two-level CES function to describe production in the forestry and logging industry. We can therefore, in principle, use exactly the same techniques to estimate the relevant substitution parameters. Unfortunately, there are significant data problems, and the estimates that we are able to obtain are somewhat less than satisfactory. The problem lies in separating out the contribution of natural resources from the contribution of capital. The cost shares of capital are usually determined residually, by subtracting the cost share of labour (for which data is widely available) from one. Where there are more than two primary factors, this method clearly is insufficient.

We need a method of determining either the cost share of capital or the cost share of natural resources (the remaining cost share again being determined residually). While in principle this represents no particular problem, in the case of New Zealand it raises a number of difficulties.

The first problem lies with the lack of data on New Zealand capital stocks. Because this information is not available, it is necessary to utilise information on land (as a proxy for natural resources instead). Data on land use in the forestry industry is available (from the Ministry of Forestry, 1996), but only annually. Another problem arises here because, in order to calculate cost shares, we need a price for the use of this land. The most obvious proxy for the opportunity cost of land use is the rental rate of the land, but again, such data is not available. We are forced to take a more indirect approach to imputing the price of land use. Data on the average sale price of all rural land is available from Valuation New Zealand (1996). By accounting for the sales of farming land, we are able to separate out non-farming rural land sales (unfortunately, this includes mining land and well as forestry land). Multiplying the average price per hectare of this land by the number of hectares planted, we obtain an estimate of the value of the land used in forestry. Multiplying this by the market interest rate (we use the 90 day bill rate, as above), we obtain an estimate of income forgone by holding onto the land used, which we utilise as our 'implicit' rent. This approach, while far from ideal, at least gives us some indication of the opportunity cost of land holdings. However, the necessary data is available only from 1983. Moreover, data on salaries and wages paid to labour, output of the forestry industry and purchases is also available only annually (from the Annual Enterprise Survey – Statistics New Zealand), and only from 1986 (more detailed quarterly financial data used above is available only for the manufacturing sector, not primary). This immediately limits us to only nine observations, which is clearly not sufficient for us to have any confidence in our estimations. The cost of capital is again the interest rate on 90 day bank bills, while the we use the Labour Cost Index (Statistics New Zealand) for the forestry and logging category as our labour price. Bearing the severe limitations of the data in mind, the results of estimation of the CRS restricted regression using the same procedure as above. We estimate on the capital and land equations. The R-squared statistics for each equation are 0.28 and 0.83 respectively. The Durbin-Watson d statitistics are 0.56 and 1.75. We test the hypothesis that $\sigma_{LN} = \sigma_{KN} \neq 1$, and cannot reject (the χ^2 is 0.63). We can reject the hypothesis that $\sigma_{LN} = \sigma_{LK}$ (the χ^2 is 7.55). We therefore use the same technique as above to calculate the corresponding inner and outer elasticities for the two-level CES. Taking the average we obtain 0.87 and 0.57 for the inner and outer elasticity estimates. These figures are in the same ballpark as estimates obtained by other authors (Harrison and Kimbell, 1985, obtain figures of 0.78 and 0.64 for times-series and cross-sectional estimates of the elasticity of substitution between labour and capital in forestry, based on other econometric work), and indicate that there is more substitutability between labour and capital than between value-added and land, as we would expect. However, given the data quality, we have little confidence in the results, although we consider them plausible. We conduct substantial sensitivity analysis on these estimates.

Other Production Elasticities

All other production elasticity estimates are taken from previous studies. The New Zealand Julianne Model assumes Cobb-Douglas production functions, and the Australian ORANI model assumes that all production elasticities are 0.5. The two most obvious sources of external estimates are therefore of little value. Instead, we make use of the estimates used in the model of the Australian economy presented by Harrison and Kimbell (1985). These elasticities are mapped onto our small-scale model (using weighted averages). The results are presented in Table 6.9. Note that there are two series of estimates, those based on time-series, and those based on cross-sectional data. We interpret the cross-sectional estimates as reflecting long-run trends, while the time-series estimates are more appropriate to short to medium term analysis (in general we expect to see greater substitution, and hence higher elasticities, in the long run). Accordingly, we use the time-series estimates in our short-run model, and the cross-sectional estimates in our long-run version.[12]

Table 6.9: Elasticities of Substitution Between Primary Factors

Industry	Time-Series	Cross-Section
Agriculture, Fishing and Hunting	0.78	0.64
Forestry and Logging	0.78	0.64
Mining and Quarrying	0.11	0.50
Light Manufacturing	0.90	1.09
Heavy Manufacturing	0.57	1.12
Traded Services	0.97	0.97
Non-traded Services	0.73	0.86
Construction	0.32	0.32

Source: Harrison and Kimbell (1985)

Domestic-Import Substitution Elasticities

The model we have specified uses a CES Armington aggregation function. We therefore require as our final set of parameters information on the elasticities of substitution between domestic production and imports. Once again, we make use of existing estimates to fill these parameter values. The estimates we use are from the Australian ORANI database, with the elasticities used in this model mapped by sector to our small-scale model (using weighted averages). The elasticities used are shown in Table 6.10, and range from 1.4 to 2.0. As no other data is available, we use these elasticities in both the short and long run versions of the model.

Table 6.10: Domestic-Import Substitution Elasticities

Industry	**Elasticity**
Agriculture, Fishing and Hunting	1.7
Forestry and Logging	2.0
Mining and Quarrying	2.0
Light Manufacturing	1.7
Wood Product Manufacturing	1.7
Heavy Manufacturing	1.4
Traded Services	2.0

Source: Dixon et al. (1982)

Model Calibration and Implementation

Variable Assignment

Calibration is the process of determining the values of the model parameters that will cause the model to replicate the base-year equilibrium. We begin with a definition of units. We define the initial domestic prices of imports, domestic production, and the composites (PM_i, PD_i, P_i) and the exchange rate (*XR*) as unity. This means that all the value data in the input-output table becomes volume data, and is the simplest way to separate out prices from quantities in the data. We follow the same procedure for factor incomes. Of course, in reality, labour and capital receive different incomes depending on the industry in which they are employed. This is a reflection of the different skill levels of different labour, and different productivity levels of capital. However, in this model labour (and in the long run, capital) is homogeneous, there is no skill difference. Hence setting the initial wage rate (W) at unity is equivalent to measuring sectoral labour usage in terms of an 'efficiency labour unit'. The difference in labour skills is then accounted for in the initial relative labour intensities of each industry. Foreign prices are obtained by solving back from model equations 6.15 and 6.16 using the implicit tariff rates.

Having defined our units of measurement, it is a simple matter to assign the values from our base-year dataset (Table 6.5) to the variables of the model. Input-output coefficients (a_{ij}) are obtained by dividing each entry in the matrix (1,1:9,9) by the sum of the appropriate column. Once the input-output coefficients have been calculated, the net prices (PN_i) follow immediately from equations 6.18 and 6.19.

Production and Armington Functions

We use generalised CES functions of the form:

$$X = b\left[\sum_{i=1}^{M} \delta_i K_i^{-\rho}\right]^{-1/\rho}, \qquad (6.43)$$

where M is the number of factors of production, and the δ_i sum to one, throughout the model. The ρ parameters can be obtained from the estimates of the elasticities substitution (σ), being defined as:

$$\rho = \frac{1}{\sigma} - 1. \qquad (6.44)$$

The first step is to determine the values of the share parameters that are consistent with cost-minimising behaviour. Taking the derivatives of Q with respect to the K_i and setting them equal to the factor returns:

$$b\left[\sum_{i=1}^{M} \delta_i K_i^{-\rho}\right]^{(-1/\rho)-1} \delta_i = r_i K_i^{\rho+1} \qquad (6.45)$$

Summing both sides over i and dividing 6.45 by the resultant equation gives us:

$$\delta_i = \frac{r_i K_i^{\rho+1}}{\sum_{i=1}^{M} r_i K_i^{\rho+1}} \qquad (6.46)$$

Given the values for the ρ, and the base year values of capital and labour usage by each industry, we can easily determine the δ_i parameters from the above formulae (the returns are one by construction). The values of the shift parameters (b) can then be determined using the same data in addition to the output values and the input-output coefficients given above, by solving back from the production functions.

The two level CES functions (for forestry and wood processing) can be calibrated in much the same way, the only wrinkle being that the procedure must be broken into two steps (first determining the inner level parameters, and then the outer ones using the appropriate unit cost functions).

Since the Armington aggregation functions are also of the CES form, the calibration process for these is also the same. The estimates of the elasticities of substitution (η) are given in Table 6.19, and from these the μ parameters can be determined analogously to the ρ in 6.44. The share parameters (Δ_i) can then be determined. Given these values, the $\overline{B}$ can be solved for from model equation 6.10.

Consumption Expenditure Shares

Since we have utilised a Cobb-Douglas expenditure system, consumption of each good is a constant fraction of disposable income. These fractions (the ϕ_i) are obtained simply by dividing expenditure on each commodity (C_i) by total expenditure.

Price Normalisation Weights

The weights in the price normalisation equation are given by the sectoral output shares in the base year. Hence:

$$\Omega_i = (X_i \cdot PD_i) \big/ \textstyle\sum_{i=1}^{9} X_i \cdot PD_i \,. \tag{6.47}$$

Model Implementation

The model is implemented on the GAMS system (a FORTRAN-based modelling environment) as a non-linear model, and solved using both the MINOS5 non-linear optimiser, and CONOPT2. Further details regarding the GAMS system and its component non-linear solvers can be found in Brooke et al. (1996).

Summary and Conclusions

In this chapter we have presented and discussed the construction of a computable general equilibrium model for the New Zealand economy. While the model is, like all economic models, somewhat stylised, and strongly neoclassical in flavour, it has been aggregated to provide a focus on the key industries of interest, incorporates econometric estimates of the key relational parameters, and incorporates a significant number of the basic feedback mechanisms and sectoral linkages which are of interest to us. We consider it a useful way of formalising the essential aspects of the processing incentive issue. In the following chapter we will utilise the model for running counter-factual experiments with respect to the likely effect of processing incentives in both the short and long run.

Notes

[1] This specification allows us to specify separate elasticities of substitution between capital and labour, and between value-added and natural resources. Another possibility is the CRESH function introduced by Hanoch (1971), which allows the elasticities of substitution between all factors to vary. In our econometric estimates of the relevant elasticities (for industries 1 and 2) we find that the two-level CES is an adequate representation of the technology.

[2] Note that this imposes restrictions of zero cross price elasticities and unitary income and uncompensated own-price elasticities.

[3] The expression can be obtained by solving for the cost minimising usage of D_i from the Armington aggregation functions (6.10), and dividing through by Q_i. The actual derivations are contained in Appendix B. See Dervis et al. (1982) pp.224-5 for further details and general discussion.

[4] Once again, the derivations are relegated to Appendix B, and further details and general discussion can be found in Dervis et al. (1982) pp.222-3.

[5] Given cost minimising behaviour by users of domestic production and imported goods, the composite prices are given by the unit 'cost' functions corresponding to the CES Armington aggregation functions (see Appendix B for the derivation).

[6] In principle it is possible to normalise around virtually any nominal magnitude in the economy, but we choose to follow the system of Dervis et al. (1982, p.150) of a 'no-inflation' benchmark normalisation rule. Note that we interpret equation 6.29 purely as the choice of a numeraire. Our model remains a barter model.

[7] That the law of one price implies extreme specialisation in an economy characterised by constant returns to scale production and where the number of goods exceeds the number of factor of production was first shown by Samuelson (1953).

[8] An alternative to the assumption of long-run capital rigidities would be to incorporate the Armington assumption into the export side of the model, as well as the import side. This is the procedure followed in de Melo and Tarr (1992). This ensures that exports can never hit zero, and therefore adds limits the extent of production swings. However, while one can easily imagine differentiated imports and domestic production (for example, New Zealand produces softwood while importing hardwood), this not so clear on the export side (i.e., it is equivalent to assuming that domestic production for export differs from domestic production for domestic use – in the case of forestry this is certainly not the case). It is for this reason that we prefer the more transparent method discussed here.

[9] Where there are two factors of production, the AES is the same as the standard definition of the elasticity of substitution multiplied by a scale factor, which is equal to one with constant returns to scale technology.

[10] Note that since we do not have data on natural resource use, we are forced to assume that the elasticities are the same as between capital and the other inputs (hence the generalised CES in the second level of the production function).

[11] See Berndt and Christensen (1973) for details of the derivation of these testable conditions.

[12] We use the same estimates presented above for the wood products industry in both the short and the long run, despite the fact that they are time-series estimates. We feel that given the considerable stability we observe in the estimates over a twelve year period, this is not unreasonable.

7 Model Results

Introduction

In this chapter we evaluate various trade policy options for New Zealand forestry, using the model described in Chapter 6 to help us resolve some of the ambiguities in the results of the price-exogenous theoretical model developed in Chapter 5. We also use the model in this chapter to quantify the likely magnitude of the effects of policy intervention, something that a theoretical model cannot do. This chapter therefore serves the role of providing the quantitative information necessary for a more informed debate over the likely impact of processing incentives in New Zealand.

We present the results of our simulations by policy instrument (considering export taxes on logs, an export ban, and export and processing subsidies to wood products). We describe the short and long run impacts predicted by the model on output, trade, income distribution and overall welfare. All results in this chapter are presented for the mid-point elasticities, as discussed in the preceding chapter. We also consider the question of sensitivity of the model results to changes in key parameters. Finally, we consider two simple extensions of the model to consider income transfers to foreign owners, and environmental externalities and carbon taxes.

Export Taxes on Logs

Since it has been suggested that New Zealand should restrict its exports of logs in order to increase the level of domestic processing, we begin with simulating export taxes on logs. Export quotas have been suggested, but in a perfectly competitive model it is always possible to find an export tax that is equivalent in all respects to an export quota under the assumption that the quotas are allocated by competitive auction. Since taxes are somewhat easier to deal with than quantitative restrictions, we limit our consideration to these, but note that the equivalent quota can be found by observing the corresponding export volume. The levels of key economic variables as a percentage of the initial level with an export tax on logs are summarised in

Figure 7.1. The effect of a specific tax (20 percent) is also presented in two summary tables (7.1 and 7.2), which present percentage changes in the relevant variable from the initial equilibrium.

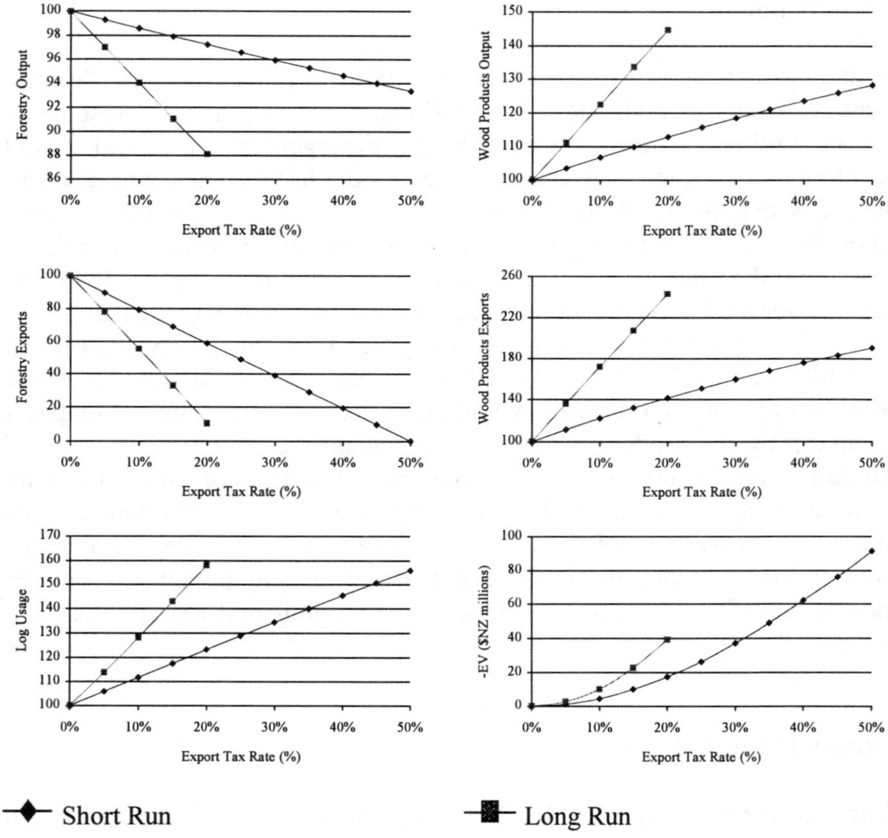

Figure 7.1: Effect of Log Export Taxes on Various Variables

Sectoral Output

The effect of the decrease in the price of forestry that the export tax causes on the forestry and wood processing industries is significant. As shown in Table 7.1, gross output of forestry in response to a 20 percent export tax is estimated to fall by roughly 3 percent in the short-run and by nearly 12 percent in the long run. The response of the wood products industry to the export tax is more substantial, a rise in output of nearly 13 percent in the short run, and 45 percent in the long run.

Table 7.1: Output and Welfare Response to Processing Incentives

Variable	Export Tax (20%)			Export Ban		Export Subsidy	
	Short Run	Long Run	Target	Short Run	Long Run	Short Run	Long Run
Forestry	-2.7	-11.9	-3.5	-6.6	-13.2	-0.0	0.0
Light Manufacturing	-0.2	-3.4	-1.0	-0.4	-3.8	-0.4	-0.9
Non-traded Services	-0.1	-0.2	-0.0	-0.3	-0.2	-0.0	-0.0
Wood Products	12.9	44.8	12.9	28.3	49.7	12.9	12.9
Heavy Manufacturing	0.2	0.6	0.2	0.3	0.7	-0.2	-0.2
Construction	-0.1	-0.0	0.0	-0.2	-0.0	-0.1	-0.0
Agriculture	-0.0	-0.1	-0.0	-0.1	-0.1	-0.1	-0.0
Mining	-0.0	0.0	0.0	-0.0	0.0	-0.0	0.0
Tradable Services	-0.1	-0.4	-0.1	-0.3	-0.5	-0.2	-0.2
EV (NZ$ millions)	**-17.5**	**-39.3**	**-3.5**	**-91.6**	**-48.2**	**-4.3**	**-1.4**

Table 7.2: Import and Export Response to Processing Incentives

Variable	Export Tax (20%)			Export Ban		Export Subsidy	
	Short Run	Long Run	Target	Short Run	Long Run	Short Run	Long Run
Imports							
Forestry	-23.3	-21.7	-6.7	-45.0	-23.3	3.3	3.3
Light Manufacturing	-0.1	-0.8	-0.2	-0.2	-0.9	-0.0	-0.2
Wood Products	3.8	9.9	3.9	6.6	10.9	8.2	5.0
Heavy Manufacturing	0.2	0.4	0.2	0.2	0.4	0.1	0.1
Agriculture	-0.2	-2.4	-0.7	-0.4	-2.6	-0.3	-0.6
Mining	0.1	0.3	0.1	0.2	0.3	-0.2	-0.1
Tradable Services	0.0	0.0	0.0	-0.0	0.0	0.0	-0.0
Exports							
Forestry	-41.0	-89.8	-25.6	-100.0	-100.0	-10.3	-8.7
Light Manufacturing	-0.4	-6.0	-1.8	-0.7	-6.6	-0.8	-1.6
Wood Products	41.5	143.1	41.4	90.8	159.0	44.8	42.3
Heavy Manufacturing	0.6	2.2	0.8	1.3	2.4	-1.3	-0.9
Agriculture	0.8	13.4	4.1	1.6	14.8	1.3	3.6
Mining	-0.4	-1.5	-0.5	-0.8	-1.6	0.8	0.6
Tradable Services	-1.1	-3.6	-1.0	-1.9	-4.0	-2.1	-1.7

The top row of Figure 7.1 illustrates the impact of increasing the level of the export tax on forestry and wood product output, respectively. Note that the long run figures end at 20 percent. This is because the export tax becomes prohibitive soon after this level, and thus the export tax ceases to have any further effect on output. Note also that gross output of wood products increases at a decreasing rate with increasing increments in the export tax on forestry, due to diminishing returns. As shown in the first two columns of Table 7.1, the response of gross output of other goods does not display such a strong pattern. As we have seen in Chapter 5, the response of gross output to a change in prices could take either sign in general. Output of all goods except heavy manufactures seems to fall in general, but the changes are relatively minor.

What is driving these results? Labour, and in the long run capital, are released from forestry production, and absorbed largely by wood processing. However, the labour and capital requirements of the extra processing are not satisfied entirely by the release of factors from employment in forestry, and hence at least some other industries experience a decline (and overall, all non-wood processing industry must decline) as factors are drawn into the processing industry, as our small-scale theoretical model indicated might be the case. Note also the increase in logs used in the wood processing industry, in Figure 7.1.

The third column of Table 7.1 is labelled 'Target'. Comparing columns 1 and 2 gives us an indication of how the impact of a given trade policy differs (in this case a 20 percent export tax on logs) over different time horizons – with differing levels of factor mobility. Another interesting question is what policy is necessary over different time horizons to achieve a given objective? In other words, taking as given that a government is prepared to use an export tax in the short run to achieve an output objective, what rate of export tax should be used to maintain that target in the long run? This is what we examine in column 3 of Table 7.1. In principle we can examine any objective and compare the policies necessary to achieve it over different time periods (export, income or employment targets, for example). However, the policy objective which seems most likely is an output target for processing. Column 3 shows the result of finding the long run export tax that maintains the same production level of wood products as column 1 (i.e., we find the long run export tax on forestry which puts the wood products industry on the same isoquant that it would be on with an export tax of 20 percent in the short run, this turns out to be in the region of 5.76 percent). Clearly, a lower export tax is required over the long run, since capital and labour are both mobile. We discuss this further below.

Trade

The response of imports and exports to a 20 percent export tax is given in Table 7.2, and in the second row of Figures 7.1. As expected, the export tax pushes down exports of forestry, and raises exports of wood products. The changes are again quite substantial. An ad-valorem export tax of 20 percent is sufficient to reduce exports of forestry by 41 percent in the short run, and by nearly 90 percent in the long run. Exports of wood products rise by 28 and 143 percent in the short and long run respectively. The pattern for imports is not quite so clear. Imports of forestry tend to fall as domestic production becomes cheaper. In general we would expect imports of other goods overall to rise as factors are drawn out of production of import competing goods and into wood processing, but which particular goods rise and which fall will depend on the system being solved. It appears that imports of all goods except agriculture and light manufactures rise slightly, but the changes are fairly minor.

Welfare Measure and Impact

Before discussing the welfare impact of an export tax, we briefly discuss the welfare measure used. In our CGE model we have a utility function as our objective function, but utility is an ordinal concept, and thus the numbers generated by this function ('utils') are not meaningful in themselves (since it is only their rank which is important). A more useful measure of the welfare changes can be obtained by using a compensation function (i.e., a money metric utility function). There are two measures of the change in utility that are commonly used, the compensating variation (CV) and the equivalent variation (EV). The compensating variation takes the post-change prices and asks what income change would be needed to compensate the consumer for the price change. The equivalent variation takes the pre-change prices and asks what income change at current prices would be equivalent to the proposed change.

In practice, both the compensating variation and the equivalent variation tend to provide fairly similar measures of the change in welfare. However, for our purposes the EV is the more suitable measure. This is for two reasons. The first is that EV measures the income change in current prices, and it is easier to judge the value of a dollar at current prices than at some hypothetical future price vector. More importantly, we will consider several policy changes (not only different levels of export taxes, but other processing incentives as well). EV is more suitable for making comparisons among a variety of projects, since it keeps the base prices fixed at initial prices (whereas with CV the base prices will be different for each policy). With a Cobb-Douglas utility function, the EV is given by:

$$EV = NDI^1 \cdot \left[\prod_{i=1}^{N} \left(\frac{P_i^0}{P_i^1} \right)^{\phi_i} \right] - NDI^0,$$

where a superscript 0 represents pre-change values, and a superscript 1 represents post change values. The EV measure is negative if welfare falls with a given policy (implying that income would have to fall at current prices to put society on the same social indifference curve), and positive if welfare rises.

Table 7.1 gives the welfare effects of a 20 percent export tax in the short and long run. We estimate the equivalent variation for the policy to be (negative) 17.5 million dollars in the short run, and (negative) 39.3 million dollars in the long run. Figure 7.1 (bottom row) displays the welfare cost as the export tax increases (the absolute value of the equivalent variation). Note that the welfare cost of a given export tax in the long run always lies above the cost of the same policy in the short run, and the increasing rate at which welfare costs increase as the export tax becomes larger (as the Harberger triangles expand).

The increasing rate at which welfare decreases combined with the decreasing rate at which output of the wood processing sector increases, implies that, as the export tax rate rises, it becomes increasing expensive in welfare terms to achieve a marginal increase wood product output by increasing the export tax rate.

Column 3 of Table 7.1 gives the long run figures for the equivalent short run output target, as described above. Once again, we observe that the absolute figures are in a sense somewhat misleading. A given policy will have a much higher welfare cost in the long than in the short run, but a given output target can be achieved at a much lower welfare cost in the long run than in the short run. The increased welfare cost for a given policy in the long run is a reflection of its increased effectiveness, but in the long run resource allocation (with capital mobile) is in fact more efficient than with the same policy in the short run. Any output target can be achieved at a lower welfare cost in the long run (or alternatively, a larger output level can be achieved at given welfare cost). The point is worth emphasising. In order to minimise welfare costs for a given output target, if export restrictions were to be imposed, despite the substantial welfare cost that the model implies they may impose on the New Zealand economy, they should be lowered over time.

Income Distribution

Many trade economists would argue that trade policy is essentially income distribution policy in disguise. We therefore now consider the distributional aspects of an export tax. This was one aspect of the theoretical model where there was some ambiguity. The top row of Figure 7.2 describes the percentage changes from the initial equilibrium in the returns to fixed and mobile factors of production with an export tax. Fixed factors in forestry (only capital is shown, but the results for natural resources are similar) see their incomes decline in the short run with the imposition of an export tax. In the theoretical model used in Chapter 5, we showed that in the long run, with capital mobile, it was theoretically possible for an export tax to cause the return to natural resources to rise (since the return to one of the mobile factors must fall). In the current model this does not occur. The changes in the incomes of mobile factors are relatively minor, due to the relatively small size of the forestry industry, and there is no possibility of them outweighing the effects of the tax on the price received. Hence, the change in the return is always negative. In fact, the decrease is larger in the long run, since there is no other fixed factor to absorb the costs imposed by the tax (we did not capture this possibility in our theoretical model since, to keep the problem simple, we ignored fixed capital in the short run).

The impact of the export tax on the fixed factors of the wood processing industry is also given in top row of Figure 7.2, and we see that the return to capital rises substantially in the short run, and the return to natural resources used in wood processing rises even more substantially in the long run.

As for the impact on mobile factors, the price of labour rises in both the long and the short run, as it is drawn into the processing industry, much as our simple model predicted. In the long run with capital mobile, we observe that the return to capital falls as the return to labour rises. This too is in line with what our simple theoretical model led us to expect, the changes are relatively small, however, and therefore so are changes to factor prices in other sectors. The export tax policy creates substantial benefits for the owners of fixed factors in the processing industry, who see their incomes rise by a considerable proportion. The incidence of the tax falls largely on fixed factors in the forestry sector. Labour gains and capital loses marginally in the long run.

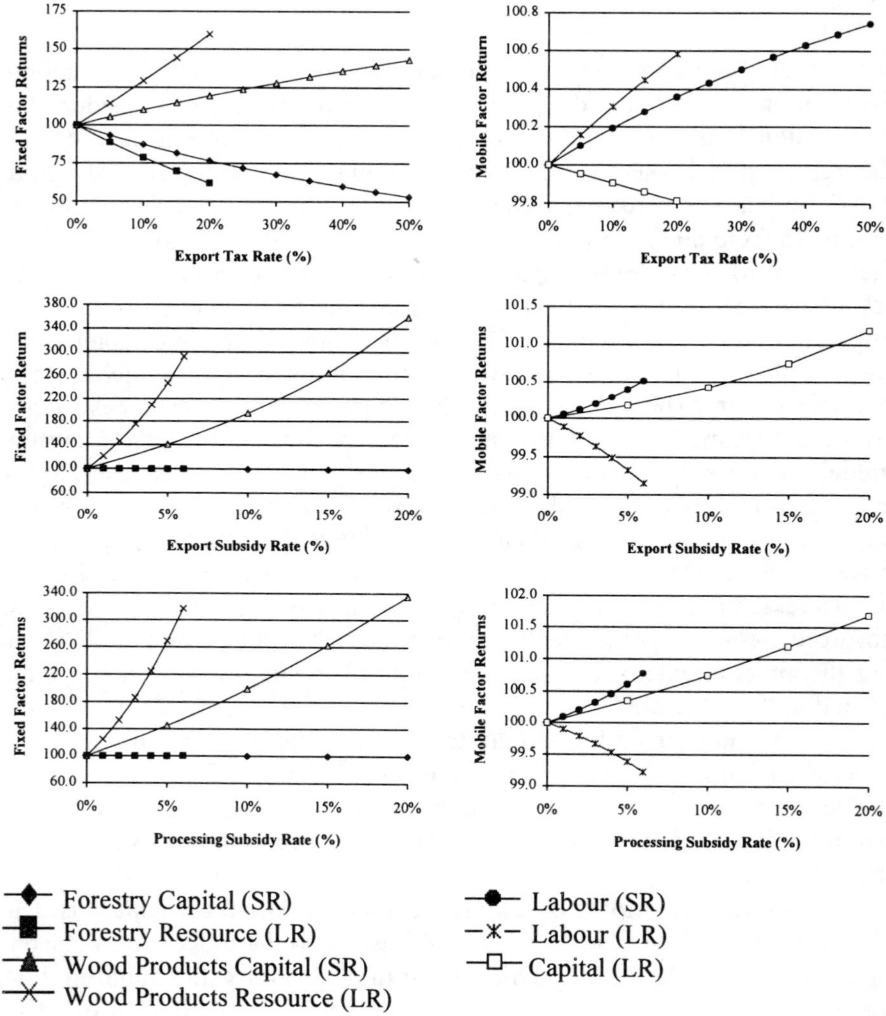

Figure 7.2: Factor Incomes Under Various Scenarios (Percentage of Initial Value)

The model results with respect for an export tax can be summarised as follows. The export tax lowers the price of forestry. This is in effect a subsidy to all domestic consumers of forestry, and hence the wood processing industry expands, drawing resources from forestry and other industries. Domestic consumption cannot absorb the increase in processed wood production, and most of the extra production is exported. Expansion of output and trade for a given level of export tax is larger in the long run, as the wood processing industry is able to choose a more efficient bundle of inputs, rather than being constrained by its initial capital stock. The policy lowers welfare for society as a whole, with higher welfare costs for a given policy in the long run (and lower welfare costs for a given objective in the long run). The export tax distorts the price vector, resulting in a sub-optimal distribution of resources in the production sector. Hence, while the production of wood products increases, the total value of production (in terms of world prices) declines. Moreover, the distorted price vector encourages inefficient usage of logs in both secondary production and final consumption – lowering welfare still further. The incidence of the export tax falls largely on the owners of those factors that cannot leave the forestry sector, who lose at the expense of fixed factors in the processing sector. Mobile factors are largely unaffected – a reflection of the size of the forestry and wood processing sectors.

An Export Ban on Logs

A more extreme form of export restrictions is a complete embargo on the export of logs. While few have suggested measures this extreme in New Zealand in recent times, the policy has considerable overseas precedence. Indonesia, Canada, the United States, and parts of Malaysia have all instituted partial or complete bans on log exports. We therefore consider it worthwhile to simulate the implications of such a move for New Zealand. An export ban is equivalent to an export tax large enough to eliminate exports entirely (a prohibitive export tax). We therefore proceed by solving the model for the prohibitive export tax levels (which turn out to be around 50.1 percent in the short run, and around 22.2 percent in the long run). This involves adjusting equation 6.25 of the model such that production of forestry is constrained to equal domestic needs, i.e.,

$$X_1 = D_1,$$

and allowing the export tax rate to be a free variable (so that the solution algorithm can find the level of the export tax that is consistent with the new constrained equilibrium). The results of the simulation are presented in the fourth and fifth columns of Tables 7.1 and 7.2 above.

As an extreme export tax, the ban causes output of forestry to fall by nearly 7 percent in the short run, and by 13 percent in the long run. Gross output of wood products rises by 28 and 50 percent, respectively. Exports of wood products rise by 91 and 159 percent (with exports of forestry now constrained to zero). The welfare costs of the policy are quite substantial, the equivalent variation being (negative) $91.6 million in the short run, and (negative) $48.2 million in the long run. As before, these costs are borne largely by the owners of forestry resources, whose income falls by 52 percent in the short run, and 42 percent in the long run. Owners of capital used in forestry see their incomes fall by 47 percent in the short run. The policy benefits owners of natural resources and capital in wood processing, who see their income rise by 41 and 43 percent respectively in the short run (the income of natural resources in processing rises by 67 percent in the long run).

In summary, an export ban on logs is effectively the export tax on logs that maximises gross output of the processing sector (the prohibitive export tax). However, in addition to maximising processing, the policy has the effect of maximising the deadweight losses (welfare costs) associated with export restrictions. Despite the numerous examples in the Asia-Pacific region of log export bans, this method of encouraging processing imposes considerable welfare costs on society.

Export Subsidies to Processed Wood Products

An alternative to encouraging processing by means of restricting exports of forestry is the subsidising of wood product exports, which encourages the production of wood products by raising its price. We can implement export subsidies in our model simply by noting that an export tax is a negative subsidy, and hence we can simulate an export subsidy to wood products by imposing a negative export tax on that sector.

Perhaps the most obvious difference between an export tax and an export subsidy is that one raises revenue while the other expends it. Of course, this is not directly relevant to assessing the overall welfare impact in this case, since we have implicitly assumed that the revenue necessary to fund the subsidy is obtained in a manner that is non-distortionary and does not disfavour any particular factor.

The results of simulating an export subsidy to wood products are given in the sixth and seventh columns of Tables 7.1 and 7.2, and in Figure 7.3 (which shows the trend in the levels of key variables as the rate of the export subsidy increases, as a percentage of the initial level). Note that columns six and seven of the tables are the impact of an export subsidy that achieves the same output target as used above (i.e., the output level achieved in the short run with a 20 percent export tax). The results can thus be compared directly to those of the export tax.

From Table 7.1 we observe that, in contrast to the export tax case, the increase in wood products output that the export subsidy causes does not come at the expense of the forestry industry with an export subsidy, but rather from a tendency to draw resources from the other manufacturing sectors. Indeed, in the long run, the export subsidy actually causes forestry to expand, as more capital flows into that sector, as described in the top row of Figure 7.3 (this is the long run effect of differing capital/labour intensity requirements in the two industries, discussed in Chapter 5, although the effect is quite slight). Note also that the downward pressure on the price of forestry with an export tax encourages a log intensive production process. An export subsidy encourages more labour intensive production. The implication is that if the objective was to increase the level of labour employed in the wood processing sector, an export subsidy (or a preferably a processing subsidy, as discussed below) would be more effective than an export tax or ban on logs.

The same pattern is also reflected in the import/export response, presented in Table 7.2 and the second row of Figure 7.3. Exports of wood products rise substantially, and exports of forestry must fall (since more of the output of this sector is being processed), but the drop in exports of forestry is not as extreme as with an export tax.

The bottom row of Table 7.1 and Figure 7.3 describe the impact on overall welfare. An export subsidy is a less damaging means of achieving a given output target than an export tax in welfare terms, with welfare costs substantially below those observed with the latter policy in both the short and the long run (since the export subsidy at least targets the industry of interest, if not the objective of interest, directly). Note however, that this applies only to levels of output achievable by an export tax. It is possible to force output much higher with an export subsidy (refer to the top rows of Figures 7.1 and 7.3), but only at the expense of substantial reductions in welfare. For example, to double wood product output in the short run would impose a welfare cost of approximately $300 million.

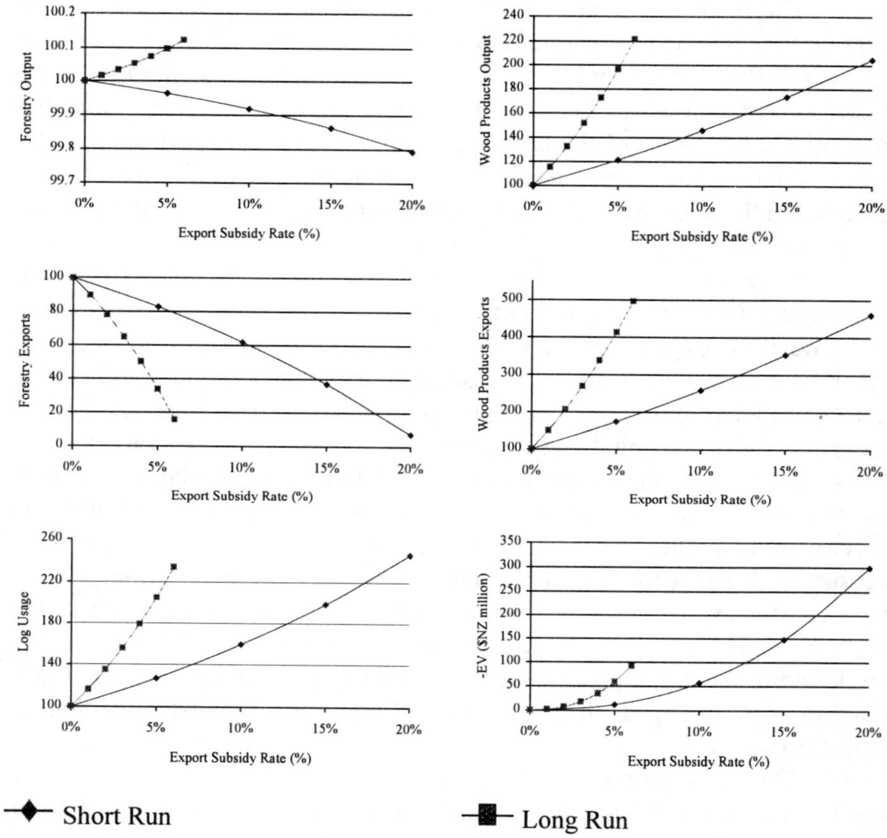

Figure 7.3: Effect of Export Subsidies to Wood Products on Various Variables

Observing the changes in factor returns reveals some interesting differences in how the cost of increased wood processing is distributed among the factors of production with an export tax and an export subsidy. The second row of Figure 7.2 gives the returns fixed and mobile factors, as described for the export tax case above. As with an export tax, in the short run an increase in processing is achieved by drawing labour from other sectors, and hence the return to labour rises. As before, the return the immobile factors employed in forestry falls as a consequence, but the effect is relatively minor (since returns are not being squeezed from the other direction by a fall in price). In the long run the results are quite different.

In the long run the owners of forestry resources experienced an even more severe drop in income with an export tax, since no other fixed factor remained to absorb the fall in price. Here, however, quite a different pattern emerges. In the long run the return to labour still rises, but the return to capital falls. Note that the return to capital falls at a faster rate than the return to labour rises. Although it is not possible to see clearly from the diagram, as the return to capital drops further, the return to forestry resources actually turns slightly upwards. This is the effect discussed in Chapter 5, where we found that the sign of the return to forest resources was ambiguous with both labour and capital mobile. The drop in the return to capital becomes sufficient to cause the return to forestry resources to rise (although the effect is very small). With an export tax this minor effect was swamped by the reduction in forestry prices. As before, the main beneficiaries of the policy are the fixed factors employed in the wood products industry.

In summary, an export subsidy is a considerably less costly means of increasing processing than restricting log exports, although it still results in an overall welfare decline for society. Of course, there may well also be other reasons to prefer a subsidy of some form if increasing processing is the objective, since they are more transparent than export restrictions. There are, however, constraints under international agreements, as discussed in Chapter 3.

Processing Subsidies to Wood Products

The final processing incentive we consider is the most direct means of influencing the level of processing, a direct subsidy to processing (output). Since processing subsidies were not included in our model as described in the preceding chapter, we make the following adjustments. First, we define the price received by producers of wood products as:

$$PS_2 = (1+s)\cdot PD_2,$$

where s is the ad-valorem processing subsidy. Next we adjust equation 6.19 (the net price equation) to reflect the subsidised price that producers now receive:

$$PN_2 = PS_2 - \sum_{j=2}^{N} a_{j2} \cdot P_j.$$

Finally, we adjust equation 6.24, to avoid double counting:

$$GDP = LI + NI + KI + TR - S,$$

where S is the cost of the subsidy (i.e., $S = PD_2 \cdot X_2 \cdot s$). With these changes in place we can now consider the impact of processing subsidies in the same manner as we have for policies that directly impact on the volume of trade above.

The results are presented in the first two columns of Tables 7.3 and 7.4 and Figure 7.4. Once again, the tabled results are for our given output target, and thus can be compared directly with the results for an export subsidy and an export tax.

Table 7.3: Output and Welfare Response to Processing Incentives

	Production Subsidy		Export Tax (FO)		Production Subsidy (FO)	
Variable	**Short Run**	**Long Run**	**Short Run**	**Long Run**	**Short Run**	**Long Run**
Forestry	-0.0	0.0	-2.8	-3.5	-0.0	0.0
Light Manufacturing	-0.4	-0.9	-0.3	-1.1	-0.4	-0.9
Non-traded Services	-0.1	-0.0	-0.1	-0.0	-0.1	-0.0
Wood Products	12.9	12.9	12.9	12.9	12.9	12.9
Heavy Manufacturing	-0.2	-0.1	0.1	0.2	-0.2	-0.1
Construction	0.0	0.0	-0.0	0.0	0.0	0.0
Agriculture	-0.1	-0.0	-0.0	-0.0	-0.1	-0.0
Mining	0.0	0.0	-0.0	0.0	-0.0	0.0
Tradable Services	-0.2	-0.2	-0.2	-0.1	-0.2	-0.2
EV (NZ$ millions)	**-4.2**	**-1.2**	**38.5**	**23.8**	**-4.0**	**-1.3**

In terms of the effect on gross output, a processing subsidy and an export subsidy seem to have much the same effect. The processing subsidy encourages exports of wood products by a lesser amount, however (Table 7.4). A processing subsidy raises the price that producers receive, but unlike an export subsidy, leaves the price that consumers pay unchanged. Hence, domestic consumption of wood products remains largely unaffected by the subsidy, and exports increase by a lesser amount (see also the second row of Figure 7.4). This fact also underlies the lower welfare cost of the processing subsidy ($4.2 million in the short run, and $1.2 million in the long run for our target). An export subsidy raises the price to consumers as well as the price to producers, it is in effect a combination of a processing subsidy and a consumption tax. Since only the processing subsidy is necessary to increase production, the consumption tax component imposes an unnecessary by-product distortion. Hence, a production subsidy results in a lower welfare cost than either an export subsidy (which targets the industry directly, but not the objective) or an export tax on forestry (which targets both the industry and the objective only indirectly), as we would expect. This is an illustration of the well-known concept of policy specificity.

Table 7.4: Import and Export Response to Processing Incentives

	Production Subsidy		Export Tax (FO)		Production Subsidy (FO)	
Variable	**Short Run**	**Long Run**	**Short Run**	**Long Run**	**Short Run**	**Long Run**
Imports						
Forestry	3.3	3.3	-23.3	-6.7	3.3	3.3
Light Manufacturing	-0.0	-0.2	-0.0	-0.2	-0.0	-0.2
Wood Products	3.9	3.9	3.9	3.9	3.9	3.9
Heavy Manufacturing	0.1	0.1	0.2	0.2	0.1	0.1
Agriculture	-0.3	-0.6	-0.2	-0.8	-0.3	-0.6
Mining	-0.1	-0.1	0.1	0.1	-0.1	-0.1
Tradable Services	0.0	-0.0	0.0	0.0	0.0	-0.0
Exports						
Forestry	-10.3	-8.7	-41.3	-25.6	-10.3	-8.7
Light Manufacturing	-0.8	-1.6	-0.5	-2.0	-0.8	-1.6
Wood Products	41.4	41.4	41.4	41.3	41.4	41.4
Heavy Manufacturing	-1.1	-0.7	0.3	0.7	-1.1	-0.7
Agriculture	1.3	3.5	0.9	4.3	1.3	3.5
Mining	0.6	0.4	-0.4	-0.5	0.6	0.4
Tradable Services	-2.0	-1.8	-1.5	-1.2	-2.0	-1.8

The impact of a processing subsidy on factor returns is described in the last row of Figure 7.2. The pattern is much the same as with an export subsidy.

In summary, the processing subsidy is the least costly means of increasing the level of wood processing taking place within the New Zealand economy. The reason for this is that it targets the objective directly. Note however, that it still imposes an overall welfare cost on the economy.

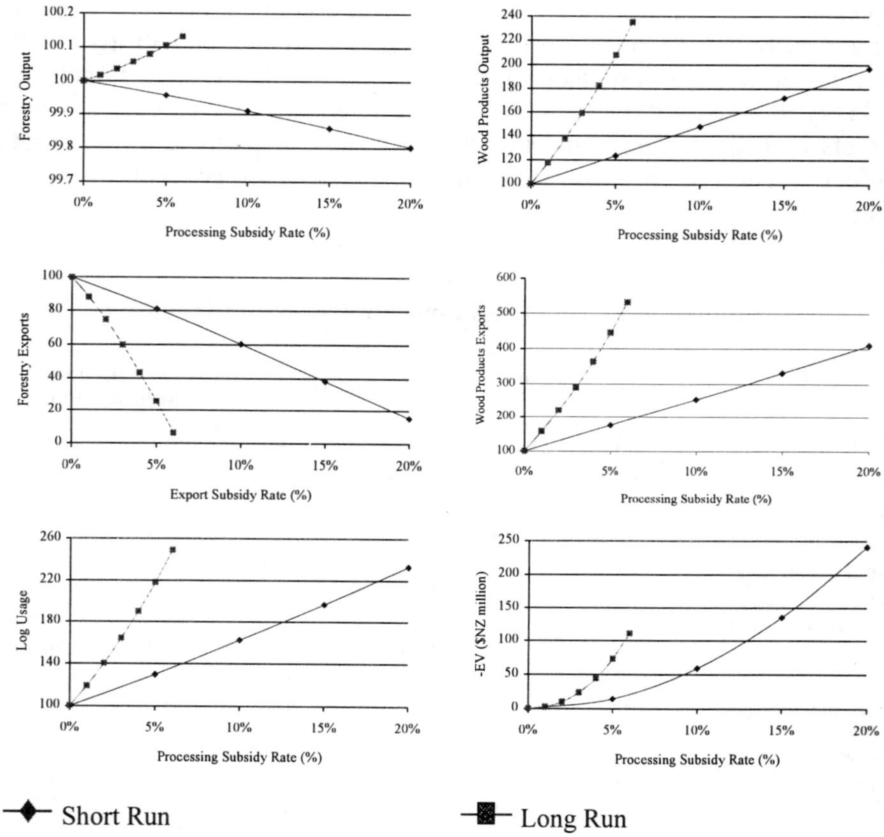

Figure 7.4: Effect of Processing Subsidies to Wood Products on Various Variables

Sensitivity Analysis

In the preceding chapter we specified a set of key elasticity parameters that we have used as the basis of our counter-factual simulation results. A key issue with models of this type is the sensitivity or robustness of the model predictions with respect to changes in the underlying parameters. To address this issue we perform what is sometimes referred to as 'conditional' sensitivity analysis. This is a series of simulations in which each parameter is separately perturbed from its central value conditional on all the other parameters remaining at their central values. The robustness of the model results is then revealed by comparison of the simulation results with the central case.

Even in a highly aggregated model like this, there are large number of parameters (the elasticities of substitution in production and the Armington aggregation functions). We focus our sensitivity analysis on the elasticities of substitution in production of forestry and wood products, since these are the sectors of most interest to us. In both the forestry and the wood products industry we have two elasticities. The internal elasticity in forestry reflects the degree of substitutability between capital and labour. The external elasticity reflects the degree of substitutability between the capital/labor composite and natural resources. Similarly, in the wood products industry the internal elasticity is between capital, land and labour, and the external elasticity between value-added and logs (the input from the forestry industry). We perturb each of these elasticities in increments between plus and minus 50 percent of their initial value in both the short and the long run, for a given policy. In the short run we use a 20 percent export tax on logs, in the long run, a 10 percent export tax on logs. We examine the sensitivity of forestry output, wood products output, forestry and wood products exports, log usage in the wood products industry, and overall welfare.

The results of the sensitivity analysis are presented in Figures 7.5 and 7.6 for the short-run and long-run, respectively. We begin with welfare, using the equivalent variation measure. In the short run the welfare measure is most sensitive to changes in the external elasticity of the wood products industry, and least sensitive to changes in the external forestry elasticity. Overall the measure seems to be quite stable to perturbations in the parameters, however. The long run simulations are somewhat more sensitive (although they are completely insensitive to changes in the internal forestry elasticity).

Gross output of forestry seems to be a little sensitive to its own internal elasticity in the short run, but insensitive to changes in the wood products technology. In the long run the only parameter that seems to have any influence is the external elasticity, which is not surprising since this involves a fixed factor. This illustrates the impact of natural resource immobility, the output changes would have been larger had more factor mobility been allowed. We observe a similar pattern in the wood products industry. In the short run the results are sensitive to the own internal elasticity, but insensitive to technology in forestry, while in the long run only the degree of substitutability between primary factors has any effect.

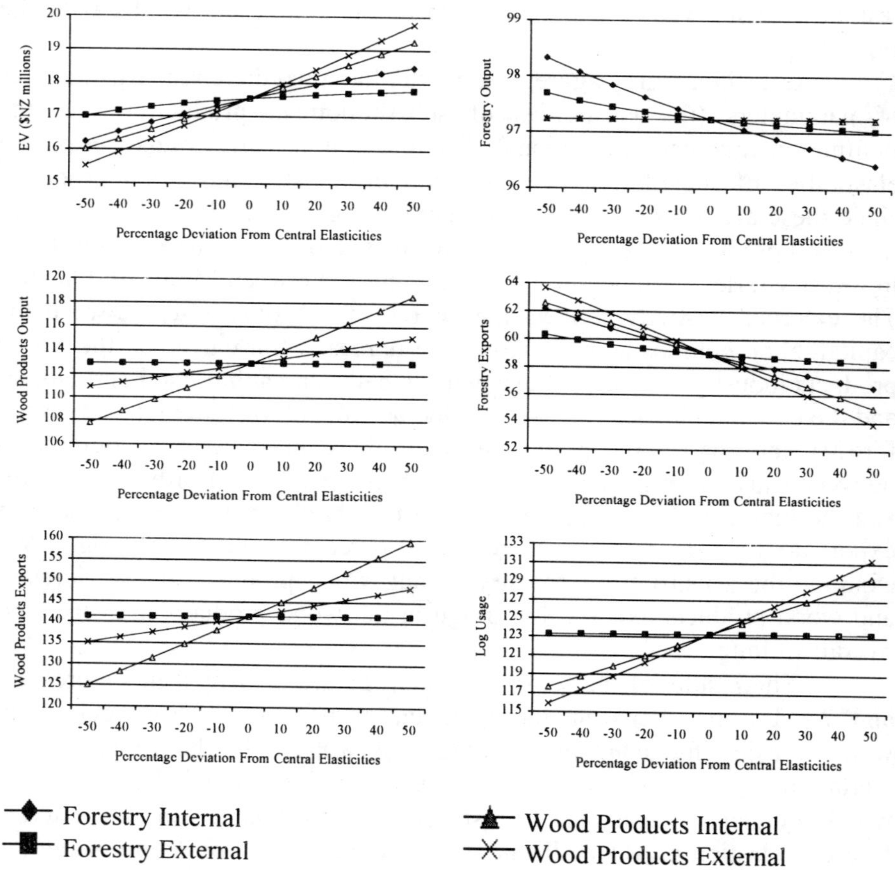

Figure 7.5: Sensitivity of Model Results in the Short Run

Exports of forestry products seem to be most affected by changes in the wood products industry in the short and long run. Similarly, exports of wood products are completely insensitive to changes in the forestry industry in both the short and the long run. Finally, log usage in wood processing. Once again, changes in the forestry sector have little impact in the short or long run. In the short run it is the ability to substitute the now cheaper logs for other factors which influences the degree of log use most (not surprisingly). In the long run the degree of substitutability between primary factors becomes more important.

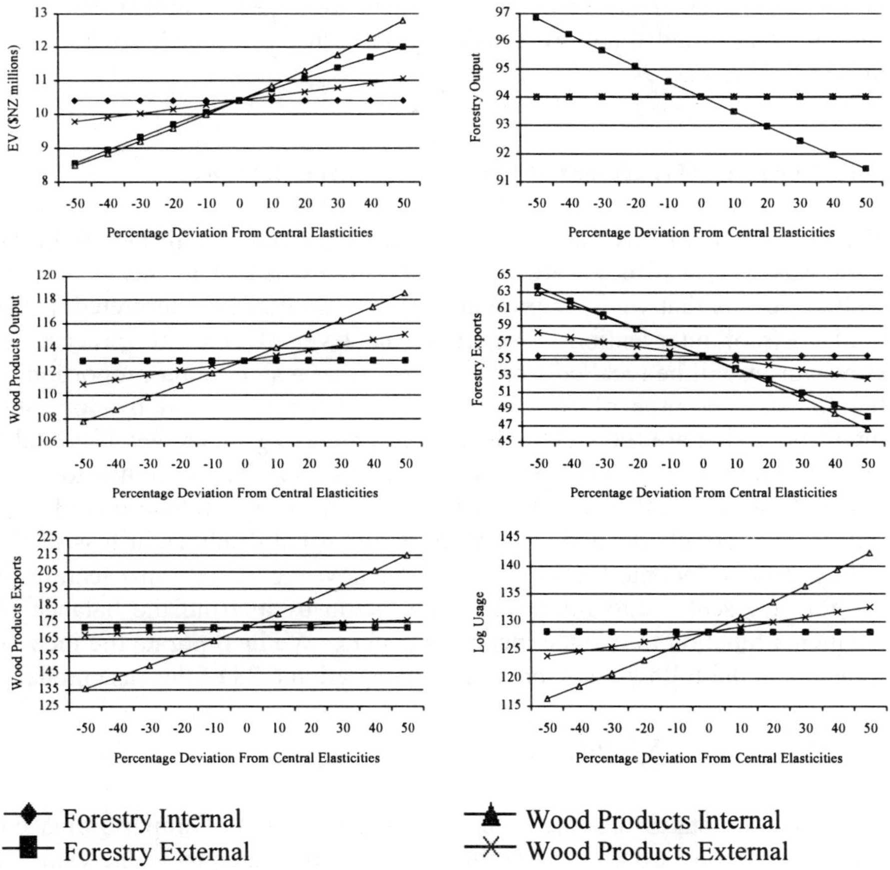

Figure 7.6: Sensitivity of Model Results in the Long Run

In summary, the model does not appear to be overly sensitive to the parameters used. As might be expected, we observe more sensitivity in the long run than the short run. Also note that most of the results (with the obvious exception of gross output of forestry), in particular the welfare cost estimates, do not seem to be particularly sensitive to the forestry elasticities used. Since these were the estimates that we had least confidence in, this makes us feel considerably more comfortable about the robustness of the results. The results do seem to be a little more sensitive to changes in the wood products industry parameters. However, the deviations are not that great (in particular with respect to the welfare cost estimates), and moreover, our estimates of these parameters were quite stable around their central values, and therefore sensitivity is perhaps not such an issue for these parameters.

An Alternative Treatment of Natural Resource Returns

The model that we have used thus far treats the returns to all factors in the same way, i.e., as if they were all domestically owned. This means that the welfare results that we have presented above can best be interpreted as the total costs of interventions to increase the level of domestic processing. However, it will be recalled from Chapter 2 that a significant proportion of the forestry resource of New Zealand is no longer owned by domestic interests. What are the welfare effects of processing incentives if we take the difference in ownership into account, and consider the return to New Zealand interests only? As we have seen from our theoretical model, this implies two effects on 'domestic' welfare, the usual deadweight loss of the policy, and an income transfer effect. We now incorporate this feature of the forestry sector into the model. We begin by altering the balance of payments function to include income transfers. We incorporate the income transfers in the following manner. Recall equation 6.28 of the base model:

$$\sum_{i=1}^{T} \overline{PW_i} \cdot M_i + \overline{F} = \sum_{i-1}^{T} \overline{PWE_i} \cdot E_i .$$

This equation states that the value (in foreign prices) of imports plus the exogenous trade surplus (defined in foreign dollars) is equal to the value of exports (again, in foreign prices). We now redefine $\overline{F}$ such that it is equal to an exogenous component, and a component that depends on the income transfers of the owners of the forestry resource, i.e.,

$$\overline{F} = \overline{F}' + \frac{n_1 \cdot \overline{N}_1^*}{XR},$$

where n_1 is the return to natural resources in forestry, and N_1^* is the quantity of this natural resource which is owned by foreign interests. Note that we have to divide through by the exchange rate because the returns to natural resources are defined in domestic currency. Since by equation 6.25 of the model, net domestic income is defined as:

$$NDI = GDP - \overline{F} \cdot XR,$$

and this forms the budget constraint of the economy, it is clear that changes in the returns to foreign owners will now affect the budget constraint faced by the economy, and hence the welfare impacts of a given policy. Note that since income transfers to foreign owners (once investment has taken place) are essentially a microeconomic phenomenon, we are not using them to explain the existence or otherwise of the current account surplus. That is to say, transfers of income derived from direct investments to foreigners can take place regardless of the initial overall level of the current account deficit/surplus (which is a macroeconomic phenomenon). We merely adjust the level of $\overline{F}'$ to be consistent with the income transfer and the initial current account deficit/surplus (although the level of the deficit/surplus will of course change as a consequence of changes in factor returns).

It remains to separate the New Zealand income from forestry from the foreign income. Data on repatriation is not directly relevant, since we believe it is the control of the income stream that is relevant, rather than whether it actually leaves the country in any given year or not (hence the model treatment of the flows going overseas is merely a stylised representation of the loss of the income stream). Roche (1992) estimated the proportion of foreign ownership of the New Zealand forestry resource at 36 percent. We estimate the figure to now be closer to 48 percent. Since our dataset is from 1993, we use Roche's figure. In our base equilibrium, with all factor returns and the exchange rate set at one, the proportion of the return to foreign forest resource owners is then calculated at \$211 million (using the input-output figures provided in Chapter 6). This is our value for N_1^*.

The results of experiments under this alternative specification are presented in the last four columns of Tables 7.3 and 7.4 (labelled 'FO' for 'foreign ownership'), and Figure 7.7. We consider first the impact of an export tax designed to achieve our given output target in the short and long run. The effects of the policy on outputs and trade are largely unchanged. However, the welfare impact of the tax is an increase of an estimated \$38.5 million in the short run, and \$23.8 million in the long run. The welfare gain as the tax rate increases is illustrated in Figure 7.7 (left). Welfare rises as the export tax rises, since the return to owners of natural resources falls, resulting in lower transfers out of the economy. It rises at a decreasing rate however, as the deadweight losses become larger, peaking where the marginal gain from income transfers exactly equals the marginal deadweight loss, and thereafter declining. The 'optimal' export tax can be calculated, and is given by 26.3 percent in the short run, and 17.4 percent in the long run. The corresponding welfare gains are \$40.3 and \$42 million, respectively.

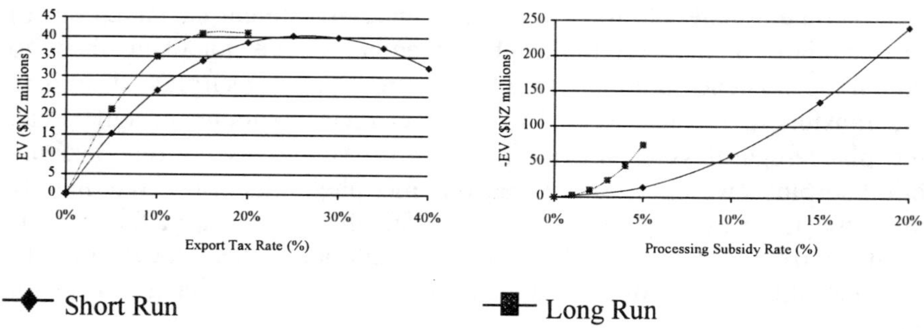

Figure 7.7: Welfare Impact of Processing Incentives with Foreign Ownership of the Forest Resource

For a processing subsidy with foreign ownership the effect of the subsidy on outputs and trade is largely the same as before. However, the welfare effects are quite different. In the short run, since the return to owners of the forestry resource declines slightly, the welfare effect of the subsidy is slightly less severe (although still negative, the changes in this factor return are nowhere near as large as with an export tax, and hence not large enough to offset the deadweight loss). In the long run, we observe exactly the opposite. The return to owners of the forestry resource actually rises slightly, as discussed above, increasing income transfers out of the economy, and lowering welfare slightly more than in our standard analysis. The results are presented graphically in Figure 7.7 (right).

These results imply that if we are only interested in the welfare of the owners of domestic factors, and if the income transfer effect is taken into account, it is welfare improving to tax exports of logs, despite the fact that this is a small country. Moreover, such a policy is significantly superior to a processing subsidy for a given processing objective. The 'specificity' rule has been turned on its head by the incorporation of foreign owned factors of production into the model. In fact, since the welfare change is still positive even at the prohibitive level in this case (although this clearly need not necessarily be the case), even an export ban on logs would be significantly superior to a processing subsidy. Once again, however, we emphasise two points. The first is such a policy is essentially nationalisation (although the method by which it takes place is slightly less transparent, and less efficient in welfare terms, than simply grabbing the assets back again outright). Second, such a policy lowers welfare for the world as a whole. Like all 'optimal' policies in the neo-classical trade paradigm, we are talking about welfare gains by one group at the expense of another, and a lowering of overall welfare. The usual disclaimer about retaliation therefore applies, although it is unclear exactly what form such retaliation might take in this case. If the foreign owners withdraw their investment, they must sell it to domestic interests (if they sell it to other foreign interests nothing changes). However, if domestic interests are only willing to pay the discounted sum of future returns, and are aware of and take into account the export tax, then the return to foreign owners is effectively transferred to domestic interests as before. Foreign investors could withhold future investment, thus forcing New Zealand to lower its current account deficit and hence lowering welfare. In the long run however, this must happen anyway – New Zealand cannot stay in a state of current account disequilibrium indefinitely. Another possibility is to increase tariffs on the processed good, pushing the price of processed good down and (possibly) raising the return to foreign factors (see Chapter 5).

This specification of the model illustrates the nature of an important problem. The standard policy recommendation to achieve a given processing objective should be subject to considerable case-by-case scrutiny, since changes in the ownership of resources can lead to counter-intuitive results. This is clearly important in the case of New Zealand forestry, and may also be of importance for developing countries wishing to encourage domestic processing of their resources, where those resources are partially owned or controlled by foreign companies.

Environmental Externalities

While in New Zealand the main incentive for the proponents of log export restrictions has clearly been the expansion of domestic processing, proponents here and overseas have also been able to argue that export restrictions are good for the environment. Perhaps the biggest environmental issue facing the global forestry industry is the unsustainable extraction of roundwood from forests to supply international trade in forest products, which can lead not only to a decline in the availability of forest resources, but also wider environmental effects. These externalities include the loss of the resource for other consumptive uses, the loss of ecological functions and the loss of other non-consumptive values of the forest. It may be argued that export restrictions are helpful because falling external demand may reduce log harvests. Moreover, once the processing industries expand and become dependent on a regular wood supply, incentives to investment in reafforestation may expand.

Although this is clearly a complex issue, OECD (1994) argues that the current evidence does not support these arguments; log export bans have led to neither better forest conservation nor to the development of efficient processing industries, since the bans do not reduce the overall demand for logs, but rather shift the location of processing. While log export restrictions may stimulate short-term growth of domestic processing, over time they tend to result in undervaluing of logs, processing over-capacity and inefficient production processes. Barbier et al. (1995) similarly, does not find support for the hypothesis that export restrictions lower the rate of deforestation. Moreover, some of the debate over the environmental impact of log export restrictions appears to be somewhat out of focus, taking as given the objective of reducing the rate of deforestation. The environmental cost associated with forest products trade is not deforestation itself, but rather the impact which deforestation has on the environment. Although they are difficult to measure and quantify, it is these costs that result from deforestation which are the objective of economic analysis, rather than the deforestation itself.

In New Zealand, these consequential costs of deforestation give the environmental argument for log export restrictions a further spin. As discussed above, the forest resource available for economic exploitation in New Zealand is not a natural forest, as is the case in most tropical hardwood suppliers, but rather commercial plantations of exotic softwood species. Log export restrictions are likely to result in lower levels of forestry output. However, as these forests are largely privately owned and planted, this reduction in output may not imply a reduction in deforestation, but rather a reduction in reafforestation.

While the role of plantation forests in the ecosystem may be a subject of debate, there is no debate over the role of plantation forestry in the New Zealand strategy for achieving its carbon emission reduction targets under the FCCC. Article 4.2 of the FCCC contains specific commitments for Annex 1 parties (of which New Zealand is one), including a target to reduce the levels of CO_2 gas emissions to 1990 levels by the year 2000. At the core of the New Zealand strategy is increased plantations of forestry to act as 'carbon sinks'. Indeed, it was initially expected that some 80 per cent of the target reduction would be met by increasing absorption through forestry, with only 20 per cent through reduced gross emissions (voluntary reductions, increased energy efficiency, etc.) The New Zealand policy package announced in 1994 also includes the imposition of carbon charges in the event that the target will not be achieved by these other means.

There are a number of areas of interest here, not the least being whether or not targeting net emissions is rational from a long term perspective (since growth rates in the forestry biomass cannot be expanded or even maintained indefinitely), or scientifically valid (measurement of absorption is difficult, and there is some debate over whether a unit of carbon absorbed really is the same as a unit not emitted – since the forest is only a store of carbon). Since log export restrictions remain an area of debate in this country, also of considerable interest is the impact that log export restrictions might have on the level of gross carbon emissions as the industrial structure is changed, on the absorptive capacity of forestry in New Zealand, and on the steps necessary to ensure that the carbon emission targets are met and their associated costs.

CGE models have already been used in numerous studies of trade and environment issues. In some cases existing models have been extended to include environmental analysis. One example of such an approach is Marks et al. (1991), who extend the ORANI model of Australia to examine the costs of carbon emission abatement. There are also numerous extensions of the GTAP model, such as Perroni and Wigle (1997), which uses side modules which calculate emission responses to various policies, using the GTAP results as inputs. Other studies use models specifically built for environmental policy analysis, including various modified versions of the GREEN model, developed at the OECD (described in Burniaux et al. 1992). In the case of New Zealand the only published CGE estimates are derived from the multi-country WEDGE model, as described in Chisholm et al. (1994).

In the remainder of this chapter we describe modifications of our model of the New Zealand economy to consider environmental issues.

Model Changes

We adjust the short-run version of our model. The input-output database is again aggregated to nine sectors, this time separating energy rather than construction from non-traded services (the trade and production data is summarised in column one of Table 7.5). While many other CGE analyses of environmental issues have maintained the standard procedure of assuming all intermediate inputs are utilised in fixed proportions to output, in our model we allow energy to be utilised by industry in variable proportions along with capital and labour (all other intermediate goods except logs are still assumed to be used in fixed proportions). The reason for taking this approach is that assuming fixed proportions in the use of energy limits the response of firms to carbon taxes (the implementation of which is discussed further below) where energy consumption is the main source of carbon emissions by industry. In a model where energy is used in fixed proportions to output the only response a firm can have to a carbon tax is to reduce its output. With flexible proportions technology, the firm can respond not only by reducing the scale of its activities, but also by substituting less pollutive inputs (capital and labour) for more pollutive ones (energy). We believe that this flexibility is an important model feature.

Unlike some other studies (Espinosa and Smith, 1995, for example), we do not include demand side effects of pollution through indirect health links or direct traffic effects, etc. The only measure of the benefit of emission policies is the reductions in emissions themselves.

We incorporate carbon emissions using 1993 New Zealand data, presented in Table 7.6. In the model emissions can come from three distinct sources. The first is from the production process itself (e.g., fugitive fuel emissions). The second is from consumption of energy products for the purpose of production. Finally, we have emissions from final consumption of energy products. For all of these sources we use fixed emission coefficients. Note that different production processes lead to different levels of emissions by level of energy consumption, reflecting the fact that different industries use different sources of energy intensively (coal, oil, electricity, etc.). We also incorporate carbon absorption into the model, which is assumed to be related by fixed coefficients to the level of gross output of the forestry sector. Carbon taxes are incorporated into the model by means of a specific tax on emissions, the rate of which is the same irrespective of the source of those emissions (hence the single carbon tax is in effect a combination of taxes on consumption, production, and inputs to the production process).

The model differs from that used by Chisholm et al. (1994) in several ways. The database is newer, and we use a slightly different production structure (in particular different sources of energy are assumed to be used in fixed proportions in each industry, and energy can be substituted at different rates for capital and labour in each industry – rather than at a constant rate across industries as in WEDGE). Emissions can come from productive processes directly, as well as energy consumption, and the model also incorporates, albeit somewhat crudely, carbon absorption.

Model Results

We consider three experiments. In the first we use the model to find the carbon tax that reduces net New Zealand carbon emissions to their 1990 level, as under the FCCC obligations. We next analyse the impact of a 25 per cent export tax on logs, which we assume is implemented for the purpose of increasing processing. Finally, we consider the same export restriction in combination with a carbon tax to meet FCCC requirements.

The results of these experiments are also contained in Tables 7.5 and 7.6. Table 7.5 gives the percentage changes in production and exports, Table 7.6 the percentage changes in energy usage by sector, carbon emissions and summary statistics. Consider first the impact of a carbon tax (column two of both tables). The carbon tax required to reduce New Zealand's carbon emissions from 12449 kt in 1993 to the 1990 level of 7772 kt is NZ$67,410 per kt of carbon emitted. This figure is broadly consistent with existing estimates. It is also a substantial tax, and not surprisingly the impact on production is substantial. All sectors except services respond to the tax by reducing output. As we might expect, it is the dirtiest industries (energy, mining and quarrying, and the manufacturing sectors) that exhibit the largest response. However, as the figures in Table 7.5 and 7.6 indicate, the production response is actually relatively small compared to the reduction in energy use by industries (see also the reduction in final energy consumption). As we might expect, all industries reduce their energy consumption substantially in response to the carbon tax. Moreover, in this model it appears that the effect of substituting energy for other factors of production dominates the production scale effect on emissions.

In Table 7.6 we have the reduction in emissions by sector. Note that since the carbon tax marginally reduces the output of the forestry sector, absorption declines slightly. The equivalent variation or welfare cost of meeting the FCCC obligation is calculated at NZ$154.6 million, or approximately 0.2 per cent of GDP. Of course, the model places no value on the reduction in carbon emissions that the policy has induced, and hence the carbon tax may in fact be resulting in a welfare gain (if society does indeed value the improvement in the environment more highly than the loss of income).

Table 7.5: Production and Trade Effects of a 25% Log Export Tax

	Initial Value	Carbon Tax	Export Tax	Carbon + Export Tax
		Percentage Change		
Production (NZ$ millions)				
Forestry	2036	-0.2	-3.7	-3.9
Light Manufacturing	19329	-6.3	-0.3	-7.4
Non Traded Services	39391	0.6	-0.2	0.5
Wood Products	2486	-3.8	16.0	11.3
Heavy Manufacturing	19965	-4.4	0.3	-4.7
Energy	8170	-11.8	-0.0	-13.5
Agriculture	10679	-1.0	-0.0	-1.2
Mining	480	-17.5	-0.4	-23.6
Tradable Services	49918	0.7	-0.2	0.7
Exports (NZ$ millions)				
Forestry	492	3.2	-52.0	-47.7
Light Manufacturing	9654	-10.9	-0.4	-12.7
Non Traded Services	0.0	-	-	-
Wood Products	672	-11.7	51.4	36.9
Heavy Manufacturing	4149	-15.4	1.0	-16.4
Energy	226	203.1	-2.1	222.2
Agriculture	1568	19.4	0.9	22.8
Mining	175	-40.1	-1.5	-55.9
Tradable Services	6171	9.6	-1.3	10.0

Consider next the effect of the imposition of an export tax on logs. The export tax lowers the domestic price of forestry, and hence production of forestry falls. As before, the policy has the desired effect on the processing industry. The lower price of forestry is effectively a subsidy to wood processing, which expands substantially. A similar pattern follows through to exports. The misallocation of resources that the policy causes is estimated to have a welfare cost of NZ$27 million.

Table 7.6: Energy Usage, Carbon Emissions and Summary Statistics

	Initial Value	Carbon Tax	Export Tax	Carbon + Export Tax
		Percentage Change		
Energy Usage ($NZmillons)				
Forestry	13	-22.8	-25.0	-44.0
Light Manufacturing	343	-25.3	-0.2	-28.4
Non Traded Services	715	-3.4	-0.1	-4.0
Wood Products	56	-23.8	19.0	-12.3
Heavy Manufacturing	617	-19.6	0.6	-21.5
Energy	3352	-21.0	-0.0	-23.7
Agriculture	312	-15.4	-0.0	-17.3
Mining	32	-22.7	-0.4	-29.0
Tradable Services	800	-14.1	-0.1	-16.0
Carbon Emissions by Sector (kt)				
Forestry	73	-22.8	-25.0	-44.0
Light Manufacturing	2604	-16.1	-0.3	-18.2
Non Traded Services	506	-3.4	-0.1	-4.0
Wood Products	418	-16.0	17.8	-3.1
Heavy Manufacturing	4407	-15.1	0.5	-16.5
Energy	7256	-20.2	-0.0	-22.8
Agriculture	1035	-15.4	-0.0	-17.3
Mining	286	-22.1	-0.4	-28.4
Tradable Services	2117	-14.1	-0.1	-16.0
Final Energy Demand	8573	-17.9	-0.2	-20.3
Summary Statistics				
Total Gross Emissions (kt)	27275	22568	27323	22017
Absorption (kt)	-14826	-14796	-14279	-14245
Net Emissions (kt)	12449	7772	13044	7772
Equivalent Variation (NZ$millions)	0	-155	-27	-222
Carbon Tax Rate (NZ$/kt)	0	67410	0	78010
Log Export Tax Rate (%)	0	0	25	25

The new wrinkle is that the export restriction also has implications for carbon emissions. The policy causes production in a relatively clean industry (forestry) to decline, while production in the relatively dirty wood processing and heavy manufacturing industries expands. Both of these industries also expand their use of energy in production, and hence carbon emissions in these two industries expands from both of these sources. The reduction in emissions from other sectors is not enough to offset this expansion, and hence gross emissions rise, albeit only slightly. Absorption by forestry also falls, resulting in an expansion of net emissions of 4.8 per cent.

Finally, consider the impact of a carbon tax to reduce net emissions to 1990 levels in the presence of the export tax. Since the export tax alters the production structure in such a way that net emissions rise, a larger carbon tax is now required to meet the FCCC obligations. The model calculates the required tax at NZ$78,010 per kt of carbon emitted, nearly 16 per cent higher than the tax required in the absence of export restrictions on forestry. The pattern we observe in production is much the same as before, although the declines in production are larger, as we expect with a larger carbon tax. The wood processing sector still expands as a result of the effective subsidy the export restrictions provide, but the expansion is offset considerably by the carbon tax. As in the case of the carbon tax alone, in addition to reducing output, all industries economise on the use of energy.

The estimated welfare cost of the carbon taxes and the export restrictions together is estimated at NZ$221 million, or 0.3 per cent of GDP. This implies that once the FCCC obligations are taken into account, the true welfare cost of export restrictions on forestry is in fact approximately NZ$66 million – more than double the welfare cost estimated without accounting for the obligations. The entire difference between the welfare cost of the carbon tax with and without the export restrictions can be interpreted as the cost of the export restrictions, since the additional cost associated with the carbon tax comes about because of the distortions in the production structure that the export restrictions cause. Since these costs need not be incurred in order to reduce emissions to 1990 levels without the export tax, the issue of whether or not the reduction in emissions is beneficial to society overall is not relevant.

Hence, our main findings can be summarised as follows. First, the carbon taxes required to meet FCCC obligations are substantial, and will result in considerable alterations to the New Zealand production structure and energy consumption patterns. Furthermore, since New Zealand's net emissions have in fact been growing steadily since 1993, the implication is that even higher taxes than estimated here will be required. If New Zealand is committed to meeting the FCCC obligations, it can therefore ill afford to implement any policy, such as export restrictions on logs, that would lower the incentive to expand the growth of plantation forestry and hence reduce carbon absorption potential. Second, because export taxes on forestry distort production in such a way that production in dirty industries expands, and production in clean industries declines, they harm the environment in the New Zealand context (contrary to the arguments that have been used in most other markets where plantation forestry is not the norm). This distortion in the production structure greatly expands the cost of export restrictions where carbon taxes are required, since a much higher carbon tax then needs to be used to reduce carbon emissions to the desired level. Hence, we have an example of how the existence of environmental externalities in the presence of emission constraints may strengthen the neo-classical case against export restrictions.

Summary and Conclusions

This chapter has been the empirical (or perhaps numerical is a more appropriate term) counterpart to the price-exogenous models of Chapter 5. The use of a CGE framework has enabled us to numerically quantify the expected results of processing incentives to wood products in the New Zealand economy. When foreign ownership is not accounted for, processing incentives of any form are shown to reduce welfare. The usual rules regarding policy specificity apply, and processing subsidies are less damaging to the economy in welfare terms than export restrictions, which impose considerable welfare costs. The cost of a given policy is shown to be substantially higher in the long run, although an objective can be achieved at lower cost in the long run if the policy is adjusted accordingly. Where we account for foreign ownership the results are somewhat different due to the income transfer effect, and export taxes are shown to increase the welfare of domestic interests (at the expense of foreigners). The introduction of environmental externalities also alters the conclusions of the model. In the following chapter we implement our processing incentive simulations, and simulate the effects of liberalisation by other countries, in a global trade model.

8 Global Trade Analysis

Introduction

In this chapter we make use of a global trade model known as GTAP (the Global Trade Analysis Project) in our analysis of the trade policy issues surrounding the New Zealand forestry industry. In Chapter 5 we developed a theoretical two country model, where the price vector was determined endogenously. In the same way as the model described in Chapter 6 and applied in Chapter 7 can be thought of as the empirical counterpart of the price exogenous models of Chapter 5, the simulations we perform with the GTAP can be thought of as the empirical counterpart of this two-country model. The reasons for using the GTAP here are the same as the reasons for using CGE in the preceding chapters. It will be recalled from Chapter 5 that the introduction of a trading equilibrium introduced considerable ambiguity to the results. In fact, the price vector was indeterminate in general, and thus numerical simulation is really the only way of obtaining answers to the question of what happens when export restrictions are imposed, taking into account a wider set of international linkages – in particular between the prices of forestry and wood products on world markets.

The GTAP provides an existing global CGE framework, and a database that is focused on the region of interest to us – the Asia-Pacific. It is a very cost-effective means of incorporating issues of international linkages into our policy analysis. In addition, the GTAP model specification is slightly different from the single country model that we used in the preceding chapter. Hence, the results we observe from running the same counter-factual simulations as in the preceding chapter can be interpreted as another form of sensitivity analysis.

Another important reason for making use of the GTAP is that there are several simulations that we would like to consider which require a global perspective. In Chapter 3 the impact of escalating tariffs in New Zealand's markets was discussed. Using the GTAP allows us to consider directly, for example, the effects of a Japanese reduction in its escalating tariff structure, or of the APEC liberalisation agenda. These are issues that cannot be readily analysed in the context of a single country model.

The chapter is organised as follows. In the following section we briefly introduce the GTAP model. The model is well-documented elsewhere, so our discussion centres on the differences between the GTAP and the model that we developed in Chapter 6, and the methods that we use to increase the comparability of the counter-factual simulations run with the two models. We then turn to simulations using the GTAP model. We first reconsider the issue of processing incentives within the global model – exploring the degree of similarity between the results obtained and those obtained earlier, and whether such incentives can impact on other countries. Next we consider the impacts of liberalisation of the wood products industry by two of New Zealand's most significant markets in the region (Japan and Korea), and finally the implications for the forestry and wood processing industries of the broader APEC liberalisation agenda. A summary and conclusions follow.

The GTAP Model

Overview

The GTAP lowers the costs of general equilibrium modelling by providing a standardised framework and database, which are available publicly. Its development partly reflects the frustration among CGE modellers that they were required to be a 'jack of all trades', meaning they were required to be familiar with general equilibrium theory, econometric analysis, data manipulation techniques, computer programming, and institutional structures. The GTAP sought to bring general equilibrium modelling within reach of non-specialists. Moreover, by providing a publicly available standardised framework and database, it was hoped that unnecessary duplication of effort could be avoided, and also that replicability of the experiments could be improved upon.

The GTAP's model structure, data, and implementation is described in considerable detail in Hertel (1997). Brockmeier (1996) also provides a useful overview of the model relationships, while the full details of the model database are discussed and presented in McDougall (1997).[1]

The basic structure of the model is much like other static CGE models. In each region of the model there is a Cobb-Douglas utility function over the expenditure of private households, expenditure by the government, and savings. Government spending is then allocated across commodities by a further Cobb-Douglas function, while household spending is described by a constant difference of elasticity function (CDE).

Households supply producers with endowment commodities, which can be specified as mobile or 'sluggish'. The production technology in the model exhibits constant returns to scale using single level CES functions. Unlike in our model, intermediate good use is assumed to be in fixed proportions in all sectors in the implementation of the model used here. Natural resources (land) are used in the agricultural industry only, and there is no substitution between primary factors and intermediate goods.

The GTAP model traces imports to specific agents in the domestic economy, resulting in separate payments from households, the government and firms. This innovation was adopted from the SALTER model, detailed in Jomini et al. (1991). Aside from this innovation, the treatment of trade in the GTAP and our model is similar, in particular both models make use of the Armington specification for imports, the difference being that it is also necessary to specify an elasticity of substitution between different sources of imports (as well as between imports and domestic production) in the GTAP.

One unique feature of the GTAP model is the introduction of two 'global sectors'. The first of these is a global transportation sector, which provides the services that account for the difference between *cif* and *fob* values of a given commodity shipped along a given route. The other is a global banking sector, which intermediates between global savings and investment (there remains no accounting for macroeconomic policy or monetary phenomena, investment does not come on-line to affect the productive capacity of industries or regions in the model – although it will have an indirect effect through final demand). The global closure of the model is neoclassical (i.e., global investment is equal to global savings), but this need not be the case for each region (hence the current account is flexible in the standard implementation of GTAP, in contrast to our model, and most other single country models).

The coverage of the GTAP database is substantial. The version used here (number 3) covers 30 regions, and 37 commodities. Of the thirty regions 20 are individual economies, while the remaining 10 are composite regions. The commodity listing leans towards agricultural products (see Table 8.1 for a full description). The base year of the dataset is 1992, although the data used to create the dataset comes from various years. Because of software limitations, for any given experiment it is generally necessary to strategically aggregate the data into less than ten regions and ten commodities, in order to solve the model on a personal computer.

Weaknesses of the GTAP model largely fall into the same categories as those described for CGE models in general in Chapter 6. With respect to the database in particular, there was considerable modification of the data supplied by many countries for consistency purposes, and some regions were constructed artificially. Nevertheless, provided the limitations of CGE analysis in general are borne in mind when interpreting the results, we do not feel that they detract seriously from the usefulness of the model.

Overall the GTAP falls into the same broad category as the model that we have utilised in the preceding chapters (that is, the static, neoclassical, constant returns to scale, trade-focused CGE) – extended to the multi-country case. Nonetheless, the GTAP model differs in a number of ways from our single country model, in particular with respect to the technology and the specification of the current account. While this means that simulations with the GTAP provide a way of testing the sensitivity of our results to the particulars of our model specification, it is nevertheless desirable to adjust the GTAP model so that the underlying assumptions match our model as closely as is practical.

The standard implementation of the GTAP treats both capital and labour as mobile, with only agricultural land immobile. For comparability with our 'short run' simulations we re-classify capital as a 'sluggish' factor of production and specify a near zero elasticity of transformation – which has the effect of making capital immobile. With respect to the long run, there is an additional problem in that the GTAP only distinguishes between land and capital in the agricultural sector, in contrast to our model which separates out natural resources for all sectors. Hence the 'long run' simulations of the GTAP are not directly comparable to ours. We therefore interpret the GTAP long run simulations as corresponding to a longer time period than ours, where natural resources (except in agriculture) as well as capital are free to move between uses.

The second step we undertake is to exogenously fix the aggregate current account balances in the GTAP model for all regions.[2] In the GTAP model this is not a necessary procedure in order to conduct valid welfare analysis. Nevertheless, in our model the current account is fixed, and we therefore use this procedure to ensure that we are comparing similar simulations.

The final step in maximising model compatibility is choosing an appropriate aggregation scheme. For our first set of experiments we choose the aggregation scheme laid out in Table 8.1. The commodity aggregation we have chosen is designed to match the aggregation used in our own

Table 8.1: Country and Commodity Aggregation Strategy of the GTAP Database

Country Aggregation

1. New Zealand (NZL)
2. Australia (AUS)
3. Japan (JPN)
4. Korea (KOR)
5. China (CHN)
6. Canada (CAN)
7. United States (USA)
8. South America (SAM)
  Argentina (ARG)
  Brazil (BRA)
  Chile (CHL)
  Rest of South America (RSM)
9. Other East Asia (OEA)
  Indonesia (IDN)
  Malaysia (MYS)
  Philippines (PHL)
  Singapore (SGP)
  Thailand (THA)
  Hong Kong (HKG)
  Taiwan (TWN)
10. Rest of World (ROW)
  India (IDI)
  Rest of South Asia (RAS)
  Mexico (MEX)
  Central America (CAM)
  European Union 12 (E_U)
  Austria, Finland and Sweden (EU3)
  EFTA (EFT)
  Central European Associates (CEA)
  Former Soviet Union (FSU)
  Middle East and North Africa (MEA)
  Sub-Saharan Africa (SSA)
  Rest of World (ROW)

Commodity Aggregation

1. Agriculture (AGR)
  Paddy rice (PDR)
  Wheat (WHT)
  Grains (GRO)
  Non-grain crops (NGC)
  Wool (WOL)
  Other livestock (OLP)
  Fishing (FSH)
2. Forestry and Logging (FOR)
3. Mining and Quarrying (MIN)
  Coal (COL)
  Oil (OIL)
  Gas (GAS)
  Other Minerals (OMN)
4. Light Manufacturing (LMF)
  Processed rice (PCR)
  Meat products (MET)
  Milk products (MIL)
  Other food products (OFP)
  Beverages and tobacco (BT)
  Textiles (TEX)
  Wearing apparel (WAP)
  Leather (LEA)
5. Wood Product Manufacturing (LUM)
6. Heavy Manufacturing (HMF)
  Pulp, paper, etc. (PPP)
  Petroleum and coal products (PC)
  Chemicals, rubber and plastics (CRP)
  Nonmetallic mineral product (NMM)
  Primary ferrous metals (IS)
  Nonferrous metals (NFM)
  Fabricated metal products (FMP)
  Transport equipment (TRN)
  Machinery and equipment (OME)
  Other manufacturing
7. Traded Services (TS)
  Trade and transport (TT)
  Other services (private) (OSP)
8. Non-traded Services (NTS)
  Electricity, water and gas (EGW)
  Other services (government) (OSG)
  Ownership of dwellings (DWE)
9. Construction (CNS)

model. The country aggregation is chosen to reflect New Zealand's major markets and major competitors. The first four countries are major and/or growing markets for New Zealand's forest products. Japan and Korea in particular are chosen because of the escalating tariff structures that these countries impose (refer to Chapter 4 for further discussion), which are of particular interest to us. The next three groups reflect the other major softwood suppliers on the American continent, which are New Zealand's major competitors. The other two groups are fairly large aggregations of countries. The other East Asia group has been separated from the rest of the world group to reflect the regional nature of the forest products industry.

Important summary data on this aggregation of the GTAP database is provided in Tables 8.2 and 8.3. Table 8.2 describes the pattern of production in the region. Table 8.3 contains export matrices for forestry and wood products, the source country being in the rows, the destination in the columns. Tables and figures used in the following sections express changes in percentage terms from this base equilibrium.

Table 8.2: Output of Commodities in the GTAP Database by Country Group ($US1992 billions)

Country	Agriculture	Forestry	Mining and Quarrying	Light Manufacturing	Wood Processing	Heavy Manufacturing	Traded Services	Non-traded Services	Construction
New Zealand	6.4	0.7	1.0	9.7	1.6	15.2	28.3	6.4	17.5
Australia	23.2	0.8	21.5	36.2	6.1	88.2	168.0	36.1	133.6
Japan	154.4	18.6	31.9	575.5	86.7	2179.3	2657.8	731.7	546.3
Korea	37.2	2.8	12.0	103.1	5.2	248.0	145.5	69.5	80.0
China	150.8	9.7	20.4	133.9	6.5	197.5	109.3	77.0	60.0
Canada	31.1	8.3	37.4	69.1	20.6	223.4	463.8	90.5	129.0
USA	204.7	30.8	127.9	624.5	119.5	2346.9	4067.1	762.0	1959.8
South America	124.1	5.9	63.9	250.3	18.8	450.2	438.3	96.7	167.7
Other East Asia	101.5	11.1	46.8	232.6	22.9	418.3	389.6	108.0	144.6
Rest of World	761.7	50.5	1144.3	1851.3	230.5	4452.8	6227.5	1406.4	3513.0

Table 8.3: Exports of Forestry and Wood Products in the GTAP Database by Country Group ($US1992 millions)

Country	New Zealand	Australia	Japan	Korea	China	Canada	USA	South America	Other East Asia	Rest of World
Forestry										
New Zealand	0	0	110	86	43	0	0	0	8	1
Australia	0	0	0	4	1	0	0	0	2	4
Japan	0	0	0	2	0	0	0	0	0	9
Korea	0	0	1	0	0	0	0	0	0	0
China	0	0	52	2	0	0	0	0	6	1
Canada	0	0	106	4	4	0	76	0	2	36
USA	0	2	1551	198	138	202	0	1	60	163
South America	0	0	17	47	0	0	2	3	0	170
Other East Asia	0	0	1252	277	178	0	1	0	396	146
Rest of World	0	1	666	149	80	0	33	14	229	2344
Wood Products										
New Zealand	0	171	79	5	3	1	19	0	22	25
Australia	12	0	371	2	4	1	8	1	18	79
Japan	1	11	0	34	39	18	264	5	181	154
Korea	0	1	134	0	11	2	32	1	27	90
China	3	19	313	26	0	27	347	3	330	320
Canada	7	57	1238	27	16	0	6139	3	62	914
USA	9	112	1440	128	44	1938	0	116	260	3206
South America	0	2	292	5	0	6	326	177	9	609
Other East Asia	15	266	2583	508	931	192	2786	10	1764	3246
Rest of World	18	158	554	63	82	181	2419	106	389	20949

Processing Incentives

In this section we reconsider the impact of log export restrictions on New Zealand forestry as a means of increasing domestic processing, using the GTAP model. Our main objective in replicating the export restriction experiments on a different model is to observe the degree of similarity between the results of the GTAP model and our own, and to see if the earlier results hold once a wider set of international linkages are taken into account. We also consider the impact of the policy on other countries in the region, with the objective of checking the effect of retaliating against escalating tariffs by imposing an export tax.

The results of the GTAP simulations are presented in Figures 8.1 to 8.3 and Tables 8.4 to 8.6. Figure 8.1 depicts the change in New Zealand gross output of forestry and wood products predicted by the GTAP model and our model (NZ), as the tax imposed by New Zealand on exports of forestry to all countries increases. The short run simulations of both models predict very similar declines in forestry output in response to the export tax. The long run GTAP simulations reveal a much steeper decline in forestry production than our long run simulations, reflecting the different underlying assumptions with respect to factor mobility as discussed above. The model results appear to be less compatible in the case of wood product output. While the signs of the results remain the same, the GTAP predicts much smaller changes in output than our model. While the difference in the long run simulations may be explained by the different factor assumptions, the difference in the short run simulations is likely to be a reflection of the technology assumed in the models in the processing industry. In our model we assume that the use of intermediate goods is in flexible proportions to primary factors. When the price of the intermediate good (logs) falls with an export tax, the processing industry is able to expand output by using a more log-intensive production process. In the GTAP model there is no substitution between primary factors and intermediate goods, and hence less scope to expand output.

Figure 8.2 describes the change in New Zealand welfare, using the equivalent variation measurement. We convert the EV measure of the GTAP model into equivalent New Zealand dollars for comparison. The welfare costs predicted by our model are somewhat higher than those of the GTAP. Also note that the GTAP simulations indicate an improvement in welfare for New Zealand with a sufficiently small export tax on logs, indicating that there is some, although very limited, market power.[3] However, as can be seen from Figure 8.1, the increase in processing that is likely to be achieved by imposing such an 'optimal' tax in the long run is very small. In a practical sense then, both models tend to indicate that using export restrictions on forestry to increase the level of wood processing by any substantial amount would lower overall welfare for New Zealand.

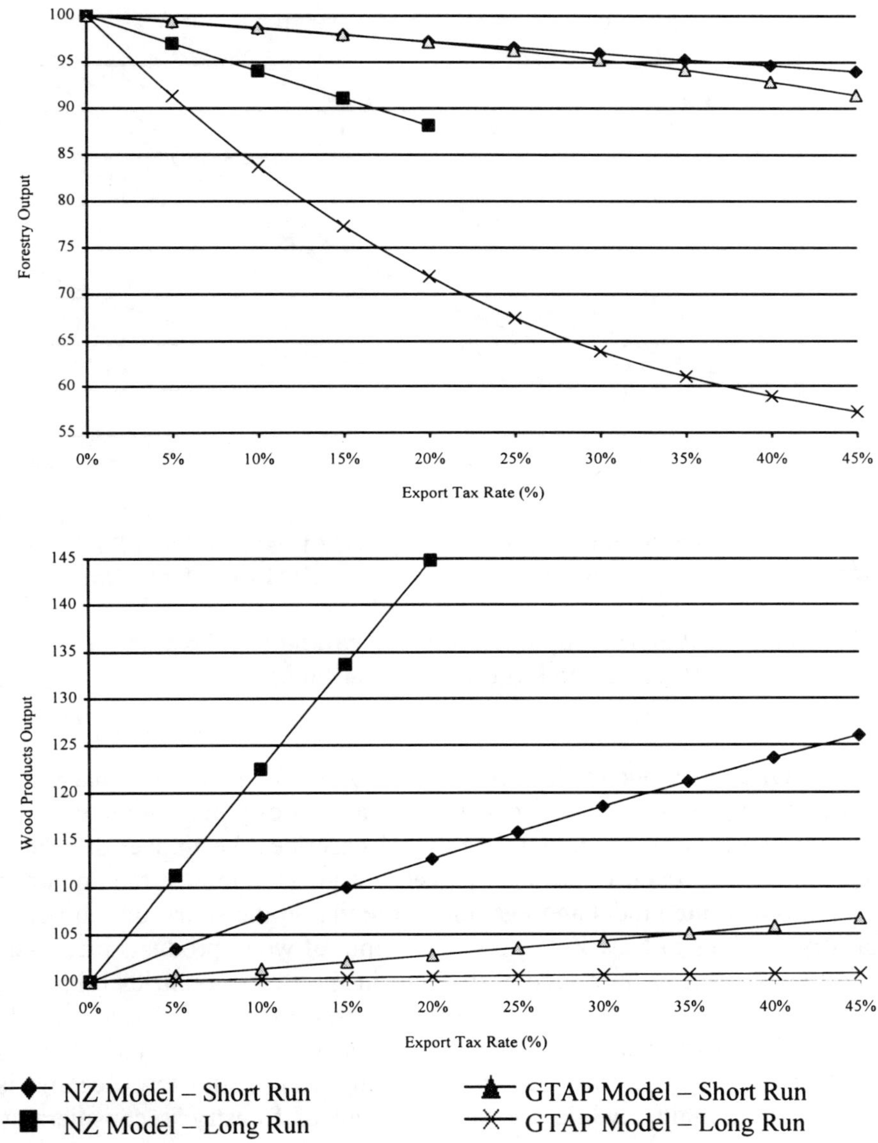

Figure 8.1: Output with a New Zealand Log Export Tax, Various Models (Percentage of Initial)

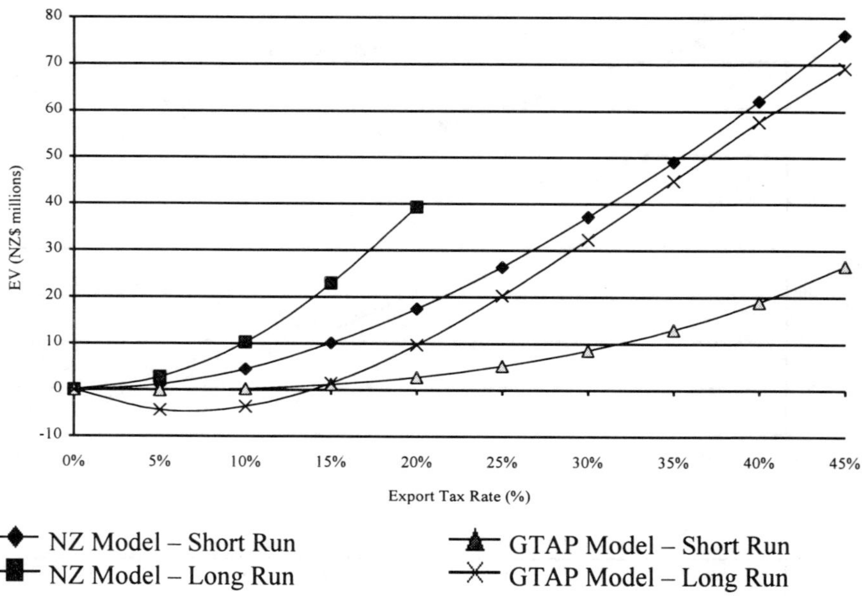

Figure 8.2: Welfare Impact of a New Zealand Log Export Tax, Various Models (NZ$1993 millions, EV)

Of course, one of the strengths of the GTAP is the fact that it is a global model, and we can thus consider the impact of policy simulations on the other regions in the model. Table 8.4 describes the impact of a New Zealand forestry export tax of 20 percent on gross output in the countries and regions of the model aggregation. Note that in the short run, output of forestry declines in New Zealand while output of wood products increases, with the opposite pattern tending to hold in the other countries (although the percentage changes are very small). An export tax does indeed provide a way of grabbing back some of the processing, albeit in a fairly limited way. In the long run, the rise in processing output for New Zealand is considerably smaller, while the fall in output of forestry is considerably larger. Moreover, we observe a slightly different pattern in the production of wood products in other countries. While output continues to fall in New Zealand's major export markets (Australia, Japan, and in particular Korea), output rises in New Zealand's competitors (Canada, the US, and South East Asia). The implication is that in the long run, the New Zealand export tax has the effect of expanding the level of processing in all the exporting countries, not only New Zealand (although forestry tends to expand by more).

Table 8.4: Percentage Changes in Output by Region with a New Zealand 20% Log Export Tax

Country	Agriculture	Forestry	Mining and Quarrying	Light Manufacturing	Wood Processing	Heavy Manufacturing	Traded Services	Non-traded Services	Construction
Short Run									
New Zealand	-0.08	-2.84	-0.03	-0.12	2.79	0.17	-0.02	-0.04	0.01
Australia	0.00	-0.03	0.00	0.00	-0.11	0.00	0.00	0.00	0.00
Japan	0.00	0.04	0.00	0.00	-0.01	0.00	0.00	0.00	0.00
Korea	0.00	0.02	0.00	0.00	-0.05	0.00	0.00	0.00	0.00
China	0.00	0.03	0.00	0.00	-0.02	0.00	0.00	0.00	0.00
Canada	0.00	0.00	0.00	0.00	-0.01	0.00	0.00	0.00	0.00
USA	0.00	0.02	0.00	0.00	-0.01	0.00	0.00	0.00	0.00
South America	0.00	0.01	0.00	0.00	0.00	0.00	0.00	0.00	0.00
Other East Asia	0.00	0.01	0.00	0.00	-0.03	0.00	0.00	0.00	0.00
Rest of World	0.00	0.01	0.00	0.00	0.00	0.00	0.00	0.00	0.00
Long Run									
New Zealand	0.53	-28.10	1.10	0.65	0.49	0.67	0.29	0.16	0.12
Australia	0.00	0.05	0.01	0.00	-0.01	0.00	0.00	0.00	0.00
Japan	0.00	0.26	0.00	0.00	-0.01	0.00	0.00	0.00	0.00
Korea	0.00	0.43	-0.01	0.00	-0.26	0.00	-0.01	-0.01	0.00
China	0.00	0.22	-0.01	-0.01	-0.01	-0.01	0.00	0.00	0.00
Canada	0.00	0.03	0.00	0.00	0.01	0.00	0.00	0.00	0.00
USA	0.00	0.20	0.00	0.00	0.00	0.00	0.00	0.00	0.00
South America	0.00	0.06	0.00	0.00	0.00	0.00	0.00	0.00	0.00
Other East Asia	0.00	0.43	-0.02	-0.01	0.00	-0.01	0.00	0.00	0.00
Rest of World	0.00	0.05	0.00	0.00	0.00	0.00	0.00	0.00	0.00

The same pattern is evident in the changes in trade. These are given in Tables 8.5 and 8.6. As before, the source regions are in the rows, the destination regions in the columns. In the short run, the shortfall created by the New Zealand export restrictions on forestry tends to be taken up by New Zealand's main competitors, Canada and South America. In the long run, all countries increase sales of forestry substantially in response to the restrictions. With respect to wood products, New Zealand increases exports at the expense of other countries in the short run. However, in the long run other economies also move to supply more wood products.

Table 8.5: Percentage Changes in Exports of Forestry by Region with a New Zealand 20% Log Export Tax

Country	New Zealand	Australia	Japan	Korea	China	Canada	USA	South America	Other East Asia	Rest of World
Short Run										
New Zealand	0.00	-8.86	-8.84	-8.16	-8.56	0.00	-9.17	0.00	-8.85	-9.04
Australia	-41.23	0.00	0.40	1.14	0.69	0.00	0.20	0.00	0.37	0.18
Japan	0.00	0.21	0.00	0.98	0.53	0.00	0.05	0.03	0.21	0.02
Korea	0.00	0.00	-0.14	0.00	0.15	0.00	-0.33	0.00	-0.16	-0.37
China	-41.33	0.18	0.23	0.98	0.00	0.00	0.04	0.00	0.21	0.01
Canada	-41.29	0.25	0.31	1.05	0.60	0.00	0.11	0.09	0.28	0.08
USA	-41.37	0.10	0.16	0.91	0.45	-0.07	0.00	-0.05	0.13	-0.06
South America	0.00	0.00	0.22	0.96	0.51	-0.02	0.02	0.00	0.19	0.00
Other East Asia	-41.43	0.00	0.05	0.79	0.34	-0.18	-0.14	-0.18	0.03	-0.17
Rest of World	-41.33	0.18	0.24	0.98	0.53	0.00	0.04	0.02	0.21	0.01
Long Run										
New Zealand	0.00	-61.63	-61.95	-59.99	-61.06	0.00	-63.01	0.00	-62.23	-62.46
Australia	-1.41	0.00	1.35	6.59	3.63	0.00	0.15	0.00	0.50	0.03
Japan	0.00	1.94	0.00	6.56	3.61	0.00	0.13	0.00	0.48	0.01
Korea	0.00	0.00	1.34	0.00	3.62	0.00	0.15	0.00	0.49	0.02
China	-1.47	1.89	1.28	6.51	0.00	-0.04	0.08	0.00	0.43	-0.04
Canada	-1.43	1.94	1.33	6.56	3.61	0.00	0.13	0.00	0.48	0.01
USA	-1.43	1.94	1.33	6.56	3.60	0.01	0.00	0.00	0.48	0.01
South America	0.00	0.00	1.33	6.56	3.61	0.01	0.13	0.00	0.48	0.01
Other East Asia	-1.45	1.92	1.30	6.54	3.58	-0.01	0.11	-0.02	0.46	-0.02
Rest of World	-1.43	1.94	1.33	6.56	3.61	0.01	0.13	0.00	0.48	0.01

Table 8.6: Percentage Changes in Exports of Wood Products by Region with a New Zealand 20% Log Export Tax

Country	New Zealand	Australia	Japan	Korea	China	Canada	USA	South America	Other East Asia	Rest of World
Short Run										
New Zealand	0.00	8.64	9.97	10.07	8.83	9.87	9.96	9.94	9.94	9.99
Australia	-3.64	0.00	0.17	0.25	0.19	0.21	0.21	0.22	0.16	0.22
Japan	-3.87	-1.26	0.00	0.01	-0.04	-0.02	-0.03	-0.02	-0.08	-0.03
Korea	-4.09	-1.48	-0.31	0.00	-0.25	-0.25	-0.26	-0.26	-0.31	-0.26
China	-3.85	-1.23	-0.05	0.03	0.00	0.00	-0.01	0.00	-0.06	0.00
Canada	-3.84	-1.23	-0.05	0.04	-0.01	0.00	0.00	0.00	-0.05	0.00
USA	-3.86	-1.24	-0.07	0.02	-0.03	-0.01	0.00	-0.01	-0.07	-0.02
South America	-3.85	-1.24	-0.06	0.03	-0.02	0.00	-0.01	0.00	-0.06	-0.01
Other East Asia	-3.84	-1.23	-0.05	0.04	-0.01	0.01	0.00	0.01	-0.05	0.00
Rest of World	-3.85	-1.24	-0.06	0.03	-0.02	0.00	-0.01	0.00	-0.06	-0.01
Long Run										
New Zealand	0.00	1.35	1.61	2.04	1.42	1.52	1.54	1.53	1.58	1.54
Australia	-0.49	0.00	0.11	0.52	0.08	0.04	0.05	0.04	0.08	0.04
Japan	-0.63	-0.29	0.00	0.38	-0.05	-0.10	-0.10	-0.10	-0.07	-0.10
Korea	-2.05	-1.69	-1.47	0.00	-1.36	-1.50	-1.51	-1.53	-1.50	-1.54
China	-0.58	-0.23	0.02	0.44	0.00	-0.04	-0.04	-0.04	-0.01	-0.04
Canada	-0.53	-0.18	0.07	0.49	0.05	0.00	0.01	0.00	0.04	0.00
USA	-0.53	-0.18	0.07	0.48	0.05	0.00	0.00	0.00	0.04	0.00
South America	-0.53	-0.18	0.07	0.49	0.05	0.01	0.01	0.00	0.04	0.01
Other East Asia	-0.57	-0.22	0.03	0.45	0.01	-0.03	-0.03	-0.03	0.00	-0.03
Rest of World	-0.53	-0.18	0.07	0.49	0.05	0.01	0.01	0.00	0.04	0.01

Changes in regional welfare are given in Figure 8.3 for selected economies in the short and long run. In both the short and the long run, the main beneficiaries of the policy appear to be the US and East Asia, while the costs seem to be borne by the main importing countries, Japan and Korea, and by New Zealand.

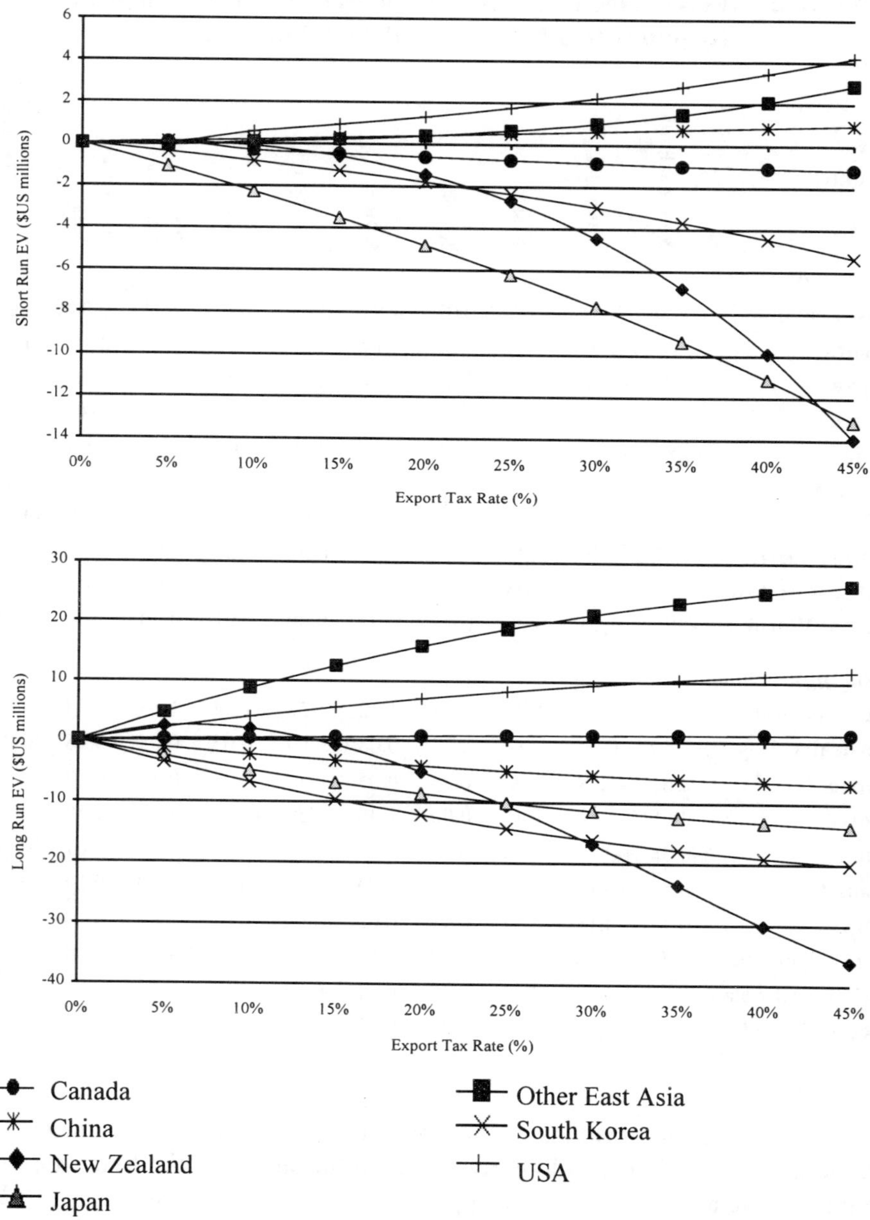

Figure 8.3: Welfare Impact of New Zealand Log Export Tax on Various Economies ($US1995 millions, EV)

In summary, the results of implementing the export restriction experiments in GTAP suggest that it is indeed possible to use export restrictions to transfer processing operations from importing countries which impose escalating tariffs to New Zealand. The strategy also lowers welfare in the target countries, so in that sense it is a 'sensible' retaliatory strategy to the escalating tariffs. However, the results indicate that it is other forestry exporting countries (New Zealand's competitors) that would benefit most from such a move. As in our model, the GTAP indicates that New Zealand would suffer considerable welfare losses as a result of implementing substantial export restrictions. Moreover, all of the output effects observed are of fairly limited magnitude, indicating that New Zealand has little influence over the price it receives in the market. This confirms the general reasonableness of our assumptions in this respect in our own model. From a political economy perspective, it also implies the scope for using such a strategy to influence our trading partners to remove escalating tariffs would be very limited.

Liberalisation by Japan and Korea

As discussed in Chapter 3, one of the most commonly cited reasons for New Zealand's tendency to export logs rather than more processed products is the escalating tariff structure of the importing countries, in particular Japan and Korea. While the main focus of this study has been the effect of utilising export restrictions or other processing incentives, in this section we use the GTAP model to round out our analysis by briefly considering the impact of the removal of these tariffs by the importing countries.

The GTAP differentiates the bilateral tariff rates applying across regions. Variation in the bilateral tariff rates applied reflects the existence of preferential trading agreements (e.g., the tariff applied by New Zealand on imports of wood products from Australia is zero, under the CER agreement) and/or the level of aggregation and the different product mix being imported from each country.

The aggregation of the GTAP used and the assumptions with respect to factor mobility and the model closure remain the same as those used above. We consider three experiments. In the first, JAPAN, Japan removes its tariffs on wood products on an MFN basis (since it is unlikely that any agreement could be reached whereby Japan removed its tariffs on a preferential basis to New Zealand). Because Japan's tariff on forestry is virtually non-existent at this level of aggregation, we do not consider the removal of forestry tariffs by Japan. In the case of Korea, we conduct two experiments. The first, KOREA1, is the removal by Korea of its tariff on wood products, again on an MFN basis. Because Korea still maintains substantial barriers to forestry in addition to its (higher) barriers to wood products, we also consider a scenario where these barriers are removed simultaneously (KOREA2). As before, we run the simulations under both short and long run factor mobility assumptions. The results are summarised in Tables 8.7 and 8.8.

Table 8.7: Percentage Changes in Output by Region Under Various Liberalisation Scenarios

	JAPAN		KOREA1		KOREA2	
	Forestry	Wood Products	Forestry	Wood Products	Forestry	Wood Products
Short Run						
New Zealand	0.01	-0.06	-0.06	0.21	0.05	0.07
	0.02	*-0.06*	*-0.07*	*0.23*	*0.06*	*-0.02*
Australia	-0.07	-0.39	-0.03	0.00	-0.02	0.00
Japan	-0.29	-0.65	-0.02	0.02	-0.01	0.01
Korea	0.04	0.35	-0.25	-4.82	-0.53	-4.15
China	0.06	0.62	-0.01	0.10	0.00	0.08
Canada	-0.02	-0.04	0.02	0.06	0.03	0.05
USA	0.00	-0.01	-0.03	0.05	0.03	0.04
South America	0.02	0.09	-0.03	0.01	0.02	0.01
Other East Asia	0.10	1.25	-0.02	0.20	-0.04	0.21
Rest of World	0.03	0.08	-0.01	0.02	-0.02	0.01
Long Run						
New Zealand	-0.13	-0.24	-0.78	0.17	1.20	0.12
Australia	-0.16	-0.63	-0.03	-0.01	-0.02	-0.02
Japan	-0.60	-0.78	0.01	0.01	0.01	0.01
Korea	0.05	0.24	-0.91	-5.41	-1.80	-4.80
China	0.02	0.40	0.00	0.10	0.00	0.08
Canada	-0.18	-0.29	0.03	0.06	0.03	0.05
USA	-0.10	-0.09	-0.02	0.05	0.11	0.05
South America	0.02	0.08	-0.04	0.01	0.07	0.01
Other East Asia	1.34	3.03	0.05	0.40	-0.10	0.37
Rest of World	0.01	0.03	-0.01	0.01	-0.02	0.01

Liberalisation by Japan

The impact of the JAPAN experiment on regional output is presented in Table 8.7. Note that there are two sets of figures for New Zealand. The upper row is the percentage change in sectoral output predicted by the GTAP model. The second row, in italics, is the percentage change is sectoral output predicted by our New Zealand model, with the price changes predicted by the GTAP fed exogenously into the model. These figures give us some insight into the similarity of the model predictions for a given change in the price vector. The changes in output predicted by the GTAP model are perhaps somewhat different to what has been predicted by the wood processing industry in New Zealand. Moreover, the figures from both models appear to match quite closely for the two sectors of interest. While wood processing does decline in Japan, as we might expect, New Zealand does not appear to benefit in terms of increased processing from the reduction in tariffs. Rather, it is the hardwood supplying countries of East Asia that appear to benefit most in this respect, in both the short and the long run. New Zealand output of processed forest products in fact declines.

The effects on overall welfare and terms of trade are given in Table 8.8, expressed in terms of the equivalent variation ($US1992 millions). Not surprisingly, it is the East Asian countries that benefit most from the Japanese liberalisation. New Zealand is largely unaffected, experiencing a slight improvement in its terms of trade and hence a gain in welfare in the short run, but a slight loss in the long run. Japanese welfare falls in line with a drop in its terms of trade, implying that the tariffs were having the effect of lowering the price of its wood product imports.

What are the factors underlying these results? Fundamentally, the substantial gain to the East Asian countries and the much smaller impact on New Zealand reflects two facts. The first is that while Japan is a major destination for New Zealand exports of forestry and wood products, New Zealand is not a major supplier in terms of the total Japanese market for either of these products. East Asia is the largest supplier of wood products to the Japanese market by a substantial margin, and it thus has the most to gain from the removal of Japanese tariffs. In this sense the results reflect the Japanese preference for hardwood species over softwood. The other important fact is that the tariff rates applied to the hardwood exporting countries of East Asia are considerably higher than those applied to New Zealand, or the other softwood exporting countries (around five times as high). Hence, while New Zealand, Canada, and the USA are relatively unaffected by the removal of the very small tariffs imposed against them, the effect on East Asia is much greater.

Table 8.8: Welfare Changes (US$1992 millions, EV) and Terms of Trade Changes (%) Under Various Liberalisation Scenarios

	JAPAN		KOREA1		KOREA2	
	EV	TOT	EV	TOT	EV	TOT
Short Run						
New Zealand	0.66	0.005	-1.00	-0.007	1.91	0.013
Australia	-8.54	-0.010	0.68	0.001	0.70	0.001
Japan	-302.82	-0.031	31.82	0.003	37.94	0.004
Korea	9.49	0.005	-111.35	-0.076	-128.55	-0.085
China	17.38	0.009	6.24	0.003	6.80	0.003
Canada	-4.29	-0.002	5.76	0.003	5.73	0.003
USA	0.22	-0.001	21.14	0.002	31.42	0.003
South America	11.73	0.007	2.07	0.001	3.80	0.002
Other East Asia	231.47	0.045	27.99	0.005	20.98	0.004
Rest of World	86.42	0.003	43.91	0.002	47.06	0.002
Long Run						
New Zealand	-0.41	-0.002	-0.42	-0.004	2.12	0.015
Australia	-5.55	-0.007	1.18	0.001	1.46	0.002
Japan	-257.10	-0.028	24.83	0.003	30.54	0.003
Korea	12.93	0.006	-95.96	-0.071	-113.23	-0.083
China	12.24	0.007	5.57	0.003	6.62	0.003
Canada	-12.69	-0.006	4.45	0.002	4.44	0.002
USA	3.32	-0.001	21.53	0.002	32.22	0.003
South America	9.93	0.006	2.49	0.001	4.45	0.002
Other East Asia	188.49	0.038	27.23	0.005	21.71	0.004
Rest of World	98.31	0.004	44.74	0.002	50.44	0.002

Liberalisation by Korea

Tables 8.7 and 8.8 also describe the results of the KOREA1 experiment (removal by Korea of its tariff on wood products on an MFN basis). The results here are perhaps closer to what the industry might expect. Output of wood products declines in Korea, and expands in most other countries, including New Zealand, in both the short and the long run (Table 8.7). The changes observed are significantly larger than those observed in the case of liberalisation by Japan. This reflects the fact that while Korea is a much smaller importer of wood products than Japan, it imposes much larger tariffs on these products.

The welfare effects of the liberalisation move are shown in Table 8.8. The results illustrate a key point which was made in our theoretical two-country model in Chapter 5. The direction of change in welfare is indeterminate, and in particular the assumption that increasing exports of wood products rather than forestry will increase welfare is not justified. In this case the GTAP predicts that New Zealand in fact loses in both the short and the long run, as its terms of trade decline. As in the Japanese case, the welfare of the liberalising country declines, again indicating that the tariffs had been having the effect of pushing import prices downwards.

New Zealand seems to have more to gain by convincing Korea to liberalise its forestry and wood processing sectors (KOREA2). In this case New Zealand experiences a slight increase in welfare. This is largely a reflection of the considerable difference in the scale of New Zealand's exports of wood products and forestry to Korea. While New Zealand accounts for only 0.6 percent of the Korean wood products market, it holds a more substantial 11 percent of the forestry market (US$4.8 million exports of wood products as compared to US$85.9 million exports of forestry). Hence the Korean tariffs on forestry, while only about one quarter of the level of those on wood products, actually have substantially more impact on New Zealand's welfare.

In summary, the effects of liberalisation by two of New Zealand's major markets as predicted by the GTAP model, are somewhat different to what has generally been thought by industry would be the case. While exports of wood products to the liberalising countries do increase and exports of forestry do decrease overall, the automatic assumption that New Zealand will benefit from this is not supported by the model (in three of the six cases considered in this section New Zealand sees its welfare decline). Moreover, in all cases the changes in output of wood products are not large. This would suggest that if New Zealand is serious about increasing domestic processing, a domestic policy such as those described in the preceding chapter would be most effective. However, such a policy cannot be justified on the basis of improving overall social welfare, which is a more sensible objective of policy intervention than increasing the level of output in one particular sector.

The APEC Liberalisation Agenda

In the preceding section we have considered the impact of the removal of trade barriers in two of New Zealand's most important forestry markets. Another interesting issue is whether or not New Zealand stands to benefit from wider ranging policy reform in the region. In this section we consider the implications of the APEC initiatives for forestry and wood products trade. The Asia Pacific Economic Cooperation (APEC) group was founded in 1989 to facilitate regional trade and to promote multilateral trade reform. Its current membership includes Malaysia, Thailand, Indonesia, the Philippines, Singapore, Brunei, China, Hong Kong, Taiwan (Chinese Taipei), South Korea, Australia, New Zealand, Japan, Papua New Guinea, the United States, Canada, Mexico, and Chile. The member countries have adopted the long-term goal of free and open trade and investment in the region, with target dates of 2010 for the developed members and 2020 for the developing economies. This represents a substantial liberalisation agenda, and the effects are of considerable importance not only for the forestry industry, but for New Zealand as a whole.[4]

The APEC initiative is unique among regional trade agreements in that it is based on the principle of 'open regionalism'. Many commentators, in particular those from non-APEC economies, have seen this as a contradiction in terms. The APEC member economies have promoted the concept as meaning encouraging intra-regional trade without discriminating against outsiders. However, even within APEC there is considerable difference of opinion as to the exact meaning of the concept, and the basis on which free trade should be established. Some countries (notably New Zealand, Australia and Japan) have argued that it means liberalisation should take place on an MFN basis, and that APEC should remain at all times consistent with the GATT Article 1 and in a supporting role to the WTO. Others countries (in particular the United States) have argued that 'without discrimination to outsiders' should mean that the benefits of liberalisation should be made available to others on the same terms and conditions as for APEC members. In other words, non-members are free to enjoy the benefits only if they reciprocate (but there is no limit on them doing so). This interpretation could mean the formation of a free trade area that is rigorously consistent with Article 24 of the GATT, and that is open to all those countries who wish to enter on the same terms as existing members. Alternatively, APEC could remain a forum for preliminary discussion only, with all actual liberalisation agreements being made under the auspices of the WTO.

Wonnacott (1995) summarises the debate, and suggests that the 'free-rider' problems that lie at the heart of the debate could be minimised if APEC were to liberalise on an MFN basis first in goods for which APEC members are the main exporters, and for which the level of intra-APEC trade is already substantial - an idea that he terms the 'predominant supplier approach'. The idea is supported by the older 'natural trading bloc' concept, where it is assumed that for countries where the level of intra-regional trade is already substantial, the trade creating effects of formation of a preferential trading arrangement are likely to outweigh trade diverting effects.

According to the GTAP3 data, in 1992 APEC country exports of forestry account for approximately 59 percent of total world exports. Moreover, of total APEC member economy exports of forestry, 93 percent goes to other APEC members. For the wood products industry the corresponding figures are 57 and 77 percent, respectively. Arguing along the predominant supplier line, these figures would tend to indicate that forestry and wood products are good candidates for MFN liberalisation by the APEC members.

The sectoral approach to APEC liberalisation was given a big push forward with the adoption of the concept of Early Voluntary Sector Liberalisation (EVSL). In 1996 members were asked to identify sectors where such early voluntary liberalisation would be beneficial. By the end of 1997 a consolidation process had led to a group of nine 'first tier' sectors, including forestry, being targeted for fast-track liberalisation as part of an Early Voluntary Sector Liberalisation (EVSL program), with a further six sectors being singled out for further investigation with a view to fast-track liberalisation in a subsequent year. A process of negotiation led to the crystallisation of a series of EVSL offers by APEC members, generally involving among other things the phasing out of tariffs in the EVSL sectors by a date well in advance of the 2010/2020 target, for those countries which opt in to the process. The negotiations ended inconclusively, with a decision to transfer these offers to the WTO, inviting other WTO members to reciprocate, leaving individual APEC members free to implement their offers unilaterally on a non-discriminatory basis if they wish.

Following this inconclusive result APEC will have to consider whether it is worthwhile to try to extend EVSL to additional sectors. Nevertheless, it is useful to consider the likely impact of sectoral liberalisation in the APEC region. In order to conduct our APEC experiments we adopt a slightly different aggregation of the GTAP database to better reflect the current membership of APEC. We leave the commodity aggregation strategy the same as above, for the same reasons. With respect to the country aggregation, the GTAP database is extremely well-suited to the analysis of APEC issues since it includes country data on all members except Brunei and Papua New Guinea. Once again, we keep New Zealand, Australia, Japan, Korea, and China as separate countries. We also separate out Chile from South America, the remaining countries moving to the Rest of World group. Because of the ten region limitation, the remaining regions are aggregates. Because of the NAFTA agreement, an aggregation of Canada, the United States and Mexico is natural. We break down the remaining APEC members into two groups. South East Asia refers to the emerging economies of the region (Indonesia, Malaysia, the Philippines and Thailand), while Other East Asia refers to those countries that are characterised by a higher level of development (Singapore, Hong Kong and Taiwan) - often referred to as the East Asian NIEs, or the 'tigers'. The final region (ROW) contains the non-APEC members.

We follow the procedure of Young and Huff (1997) in conducting our experiments, and form a post-NAFTA dataset by first eliminating barriers between the three member countries in all goods except agriculture (since the NAFTA agreement avoided liberalisation of agricultural trade). We do this so that our results are not distorted by including the effects of the NAFTA arrangement. Ideally, we would like to separate out the effects of the Uruguay Round as well, but we leave this as a topic for future research. All other assumptions remain as above.

Because the basis for liberalisation has yet to be confirmed, we follow a similar procedure to Young and Huff (1997) and Lewis et al. (1995) in conducting the liberalisation experiments for a variety of possible scenarios. These are as follows:

Experiment 1: Preferential Liberalisation: Import protection on forestry and wood products within the APEC member countries is removed, but maintained between the rest of the world and APEC. This option thus represents the creation of an APEC free trade area.

Experiment 2: MFN Liberalisation: Import protection on forestry and wood products is removed by the APEC member countries for all regions, including the ROW. The ROW does not reciprocate by lowering its trade barriers. This represents liberalisation by the APEC members on a MFN basis.

Experiment 3: MFN with ROW Reciprocation: Import barriers on forestry and wood products are removed as above, but with the ROW reciprocating on ROW-APEC trade (leaving tariffs on ROW-ROW trade unchanged).

Experiment 4: MFN with WTO Acceptance: In this experiment import barriers on forestry and wood products are removed on a global basis. This corresponds to the scenario whereby APEC implements an agreement among itself, and then takes the agreement to the WTO for ratification by all WTO members.

In all of our experiments we interpret the goal of the APEC agenda as being to eliminate tariffs completely, and the shocks that we implement in the model thus represent a move to zero tariffs for the commodities concerned. Since the actual extent of liberalisation moves is still unknown, this can be interpreted as providing an upper bound for the effects of the liberalisation agenda. We do not consider the impact of removing export restrictions or subsidies. The results of the experiments are given in Tables 8.9 and 8.10.

The impact on gross output of forestry and wood products by region is given in Table 8.9. While under the simulations we do observe a reduction in output of wood products in the major importing countries of Japan, Korea, and China, the processing is not predicted to shift to New Zealand. Under all liberalisation scenarios, New Zealand increases production of forestry, by a small proportion in the short run, and by a larger proportion in the long run. However, except under the scenarios whereby the rest of the world reciprocates with APEC in liberalising trade in forest products, New Zealand experiences a significant decrease in gross output of wood products.[5] These results would tend to confirm that relative to the other APEC members, New Zealand has a comparative advantage in forestry production, and does not have a comparative advantage in wood products. While, as discussed in Chapter 3, the proponents of export restrictions or other processing incentives may point to the lack of processed exports as prima facie evidence of the adverse effect of regional tariff structures on New Zealand's industrial structure, the statistics may rather simply reflect the fact that New Zealand is relatively good at producing forestry, and not relatively good at producing wood products. In that sense argument for processing incentives reduces to simply attempting to push production in New Zealand into an area where there is no comparative advantage – a strategy which will lower welfare for both New

Table 8.9: Percentage Changes in Output by Country Group Under APEC Liberalisation Scenarios

	Experiment 1		Experiment 2		Experiment 3		Experiment 4	
	Forestry	Wood Products	Forestry	Wood Products	Forestry	Wood Products	Forestry	Wood Products
Short Run								
New Zealand	0.22	-2.35	0.07	-4.08	0.28	0.34	0.22	-0.49
	0.25	*-1.82*	*0.08*	*-2.88*	*0.31*	*-0.42*	*0.24*	*-0.85*
Australia	-0.35	-1.97	-0.75	-3.50	0.17	0.07	-0.03	-0.62
Japan	0.30	-0.01	-0.16	-0.28	0.49	0.13	0.24	-0.01
Korea	-0.33	-2.32	-0.46	-2.91	-0.25	-0.68	-0.31	-1.27
China	-1.15	-6.12	-1.60	-7.97	-1.31	-5.56	-1.42	-6.62
NAFTA	0.11	-0.31	-0.20	-0.74	0.40	0.13	0.15	-0.31
Chile	0.25	0.41	0.15	0.00	0.90	2.97	0.79	2.19
South East Asia	0.17	2.30	-0.01	1.91	0.19	3.66	0.13	2.93
Other East Asia	3.03	6.13	1.91	4.68	4.54	9.01	3.63	7.31
Rest of World	-0.03	-0.07	0.15	0.41	-0.19	-0.83	-0.16	-0.40
Long Run								
New Zealand	2.61	-3.10	1.18	-5.00	3.15	-0.06	2.72	-0.94
Australia	-0.71	-2.99	-1.21	-4.71	-0.01	-0.60	-0.25	-1.32
Japan	-0.12	-0.23	-0.29	-0.42	-0.08	-0.21	-0.15	-0.27
Korea	-1.42	-3.17	-1.63	-3.75	-1.25	-1.53	-1.35	-2.07
China	-1.65	-10.36	-2.01	-12.49	-1.87	-10.23	-1.92	-11.25
NAFTA	-0.02	-0.39	-0.36	-0.83	0.39	0.04	0.07	-0.40
Chile	2.37	0.71	1.52	-0.37	12.99	10.36	11.20	7.63
South East Asia	3.67	7.69	1.60	5.73	5.31	12.45	4.00	9.69
Other East Asia	5.92	8.72	4.42	6.60	9.67	13.48	7.82	10.84
Rest of World	-0.19	-0.26	0.26	0.36	-0.63	-1.35	-0.48	-0.75

Zealand and the world as a whole. Indeed, the simulations all seem to quite clearly reveal that, in the event of partial or complete elimination of barriers to trade in forest products, it is Chile, South East Asia, and East Asia that see the largest increase in wood processing. The decreases experienced by

New Zealand with respect to the degree of domestic processing are much smaller with ROW reciprocation (Experiments 3 and 4), and largest with MFN liberalisation without requiring reciprocation (Experiment 2). This would suggest that New Zealand may be a relatively efficient producer of wood products when compared to non-APEC members. It certainly implies that if maintaining or expanding domestic wood processing is a priority for New Zealand, then substantial effort should be made to extend any agreement on forestry products well beyond the confines of the current APEC members.

Table 8.10: Welfare Changes (US$ millions, Equivalent Variation) and Terms of Trade Changes (%) Under APEC Liberalisation Scenarios

	Experiment 1		Experiment 2		Experiment 3		Experiment 4	
	EV	TOT	EV	TOT	EV	TOT	EV	TOT
Short Run								
New Zealand	-0.53	-0.004	-10.62	-0.073	12.21	0.073	7.87	0.045
Australia	-45.39	-0.065	-54.66	-0.098	5.40	-0.026	1.10	-0.031
Japan	29.07	0.001	34.43	0.000	84.92	0.009	166.60	0.016
Korea	-86.67	-0.064	-80.55	-0.064	-53.56	-0.049	-45.65	-0.044
China	-282.76	-0.180	-298.83	-0.210	-249.36	-0.185	-263.50	-0.192
NAFTA	-100.33	-0.007	-282.91	-0.021	157.05	0.010	21.32	0.001
Chile	9.46	0.042	2.48	-0.001	59.89	0.265	47.99	0.210
South East Asia	456.59	0.156	264.94	0.057	715.86	0.250	556.56	0.182
Other East Asia	112.77	0.046	100.95	0.040	172.02	0.070	158.23	0.064
Rest of World	62.06	0.003	666.64	0.023	-599.29	-0.023	-206.60	-0.014
Long Run								
New Zealand	-3.96	-0.024	-10.50	-0.071	6.75	0.040	4.24	0.023
Australia	-34.39	-0.053	-42.44	-0.087	14.82	-0.014	9.76	-0.020
Japan	44.18	0.002	34.87	-0.001	106.95	0.010	176.27	0.016
Korea	-67.50	-0.058	-66.93	-0.062	-34.21	-0.043	-30.45	-0.041
China	-258.41	-0.165	-267.15	-0.192	-223.52	-0.169	-234.74	-0.174
NAFTA	-96.33	-0.006	-230.49	-0.017	131.82	0.008	29.51	0.002
Chile	5.25	0.023	0.41	-0.010	53.12	0.236	42.28	0.186
South East Asia	345.41	0.109	202.40	0.030	575.11	0.191	443.35	0.135
Other East Asia	87.41	0.040	79.38	0.035	138.24	0.064	127.89	0.058
Rest of World	142.49	0.006	649.58	0.023	-465.73	-0.018	-115.12	-0.011

Table 8.10 gives the changes in regional welfare and the terms of trade, with the equivalent variation being measured in $US1992 millions. We observe that New Zealand welfare declines in the case of a preferential arrangement with other APEC members (Experiment 1), and in the case of MFN liberalisation (Experiment 2). New Zealand gains from the liberalisation move only when the ROW reciprocates.

The welfare results have some interesting political economy implications for the APEC agenda. Under a preferential agreement, the total gain to APEC members would be in the region of US$92 million in the short run and US$22 million in the long run, with most of the losses accruing to China and Korea, and most of the gains to South East Asia in both cases. Interestingly, the rest of the world actually benefits marginally from the move, its exports to the APEC members decline, forcing the price it receives on the remaining exports up. The improvement in the terms of trade increases ROW welfare slightly. By contrast, under an MFN arrangement there is a total welfare loss to the APEC members of approximately US$325 million in the short run and US$300 million in the long run, while the rest of the world benefits substantially. These results are important because they imply that the logic of the 'predominant supplier' approach does not always hold. Preferential liberalisation of the forestry products trade is shown to be substantially superior to MFN liberalisation for the APEC members, despite the predominance of APEC members in world trade in forestry products, and the extremely high levels of intra-regional trade. From the individual country results it can also be seen that all APEC members except Japan and Korea are better off under a preferential arrangement than under MFN liberalisation. Furthermore, in the case of Japan this is a short-run phenomenon, it too is better off under a preferential arrangement in the long run, and in the case of Korea the difference is only marginal. This would suggest that the interests of APEC members should coincide on this issue and a preferential arrangement should be preferred to MFN liberalisation should the ROW not agree to reciprocate. The difference in welfare for NAFTA is particularly large, a result that gives us some insight into the United States insistence on reciprocity and not allowing non-members to 'free ride' on APEC liberalisation efforts.

The other pattern that comes out clearly from the welfare results is that all APEC members experience much greater benefits if the ROW can be persuaded to join in the liberalisation of forestry products. If this occurs then all APEC members except Korea and China see their welfare improve, and total welfare of the members increases by approximately US$904 million in the short run and US$769 million in the long run (Experiment 3). These results are consistent with those obtained by Young and Huff (1997) and Lewis et al. (1995), both of whom show the gains for APEC members from liberalisation of a global basis to be substantially larger than those obtained from a preferential arrangement. Of course, the rest of the world sees its welfare decline, although the amount is fairly small when considered as a fraction of GDP, and is minimised when the ROW also liberalises amongst themselves (Experiment 4). This might raise some questions as to why the ROW would reciprocate when it is in fact better off without doing so, regardless of what the APEC members do (i.e., even if they form a preferential agreement). However, it must be recalled that we are considering the liberalisation of only one sector. The ROW would presumably gain with liberalisation of other products. Moreover, it should be noted that the ROW is not in reality some huge aggregate country, nor is it likely to behave as one. In reality there will be countries in the ROW aggregation which gain and countries which lose from liberalisation. The most important point is that for the APEC members, engagement of non-APEC members in the liberalisation process is clearly essential. The results certainly tend to suggest that a strategy of using APEC as a forum for dialogue and creating liberalisation plans that are then implemented through the WTO is a more sensible strategy than liberalising on an MFN basis without requiring reciprocation, even in goods like forestry where APEC countries are predominant suppliers. A preferential agreement, while superior to MFN without reciprocation for the APEC members, would still be substantially inferior to a reciprocal arrangement, in addition to creating obvious difficulties with respect to consistency with Article 24 of the GATT.

Summary and Conclusions

In this chapter we have made use of an existing global trade model (GTAP) to reconsider the impact of export restrictions on domestic processing, achieving results that were similar in many respects to those of the preceding chapter. In particular, the GTAP results give us little reason to suspect that New Zealand has any substantial degree of control over the price that it receives for either forestry or processed wood products, and that therefore the existence of feedback in prices, or of the possibility of using export restrictions as a strategy to force other countries to lower their trade barriers, is of little practical significance.

We have also considered liberalisation by major importing countries, and on a broader scale under the APEC agenda. Here also, the results do not support the proposition that reducing barriers to wood products trade in our major trading partners would lead to substantial increases in the level of domestic processing in New Zealand. In most of the simulations conducted it is the hardwood producing countries of South East Asia that have the most to benefit from liberalisation of forestry products, either on an individual country or a more substantial regional basis. The indications are that New Zealand's comparative advantage lies in forestry production rather than wood processing.

With respect to the APEC agenda, the GTAP results suggest that liberalisation on an MFN basis may not be the best strategy for APEC members, even in predominant supplier goods like forestry products. The results highlight clearly the importance for APEC of engagement in discussion with non-members. As was discussed in Chapter 6, CGE models have their limitations, and GTAP is no exception. The results presented here should thus be interpreted with due care. Nevertheless, they have interesting implications for the APEC process.

Notes

1 There is also a collection of information about the model and applications available at the GTAP web site: http://www.agecon.purdue.edu/gtap/.

2 This is achieved by setting the variable DTBAL(r) – the current account balance for region r as exogenous, and endogenising *saveslack(r)* – the savings slack variable. This can only be done for N-1 regions in the model, the current account in the remaining region being fixed by Walras' law. See Hertel (1997) p.53.

3 By virtue of the Armington assumption, all countries in the GTAP model will have some degree of market power. Ignoring any feedback type effects, for a small country the export demand elasticity is effectively given by the outer level Armington elasticity (i.e., the sourcing of imports elasticity). In the case of forestry this is 5.6. With respect to forestry exports, New Zealand therefore fits the GTAP definition of a 'small' country.

4 For further details on APEC see, for example, Yamazawa (1992), (1994) and (1996).

5 Figures in *italics* indicate the results of feeding the price changes predicted by the GTAP model exogenously into our own model, as above.

9 Conclusion

Summary of Findings

This study has considered the impact of trade policy intervention, in particular the application of export restrictions, in the forest products sector from the viewpoint of New Zealand, with an emphasis on the effects intervention on the level of domestic processing and welfare. The issue of restricting exports of logs in order to increase the level of domestic processing is a long-standing one in New Zealand. In this study we identified and dealt with three main arguments that have been put forward by the proponents of the use of export restrictions. The first lies in the existence of 'feedback' effects in prices. By restricting log exports it may be possible not only to raise the price of logs, but also the price of processed goods, through the price feedback mechanism. Thus it is argued that increasing domestic processing will raise social welfare. A second and related argument is that foreign countries have in effect stolen New Zealand processing by the imposition of escalating tariff structures. Using export restrictions is put forward as one way of reclaiming that processing. It may also be argued that the move may influence those countries to remove their restrictions on processed imports - with proponents of export restrictions pointing to the 'successful' example of Indonesia. Finally, we considered the fact that much of the New Zealand forestry resource is now owned by foreign interests. Export restrictions may therefore act as a mechanism for transferring income from foreign to domestic interests. The development of these debates, other possible rationales for processing incentives, and the role of the New Zealand forestry industry in the New Zealand economy were discussed in Chapters 2 and 3.

In Chapter 4 it was shown that the New Zealand literature on trade policy aspects of the forestry industry is severely lacking. In contrast to most previous studies on similar issues, which have tended to use partial equilibrium methodology, descriptive analysis, or a combination thereof, the methodology used in this study has been neo-classical general equilibrium modelling, both theoretical and applied. In Chapter 5 we developed a series of abstract general equilibrium models to examine the effects of export restrictions where an intermediate good (logs) is exported at the same time as the good into which it is processed (lumber). The chapter was intended to provide a formal general equilibrium framework for understanding the impact of export restrictions and other processing incentives on processing and welfare, and also to help to identify those areas where a formal model could not provide unambiguous indications of the effect of policy intervention, and thus to indicate where other techniques (in particular CGE) should be used. Only in the most simple abstract model of the small economy were clear formal results obtainable. In this very simple (three good) model we were able to derive unambiguously signed expressions for changes in factor incomes, resource allocation, outputs, and overall welfare with the imposition of various processing incentives. Because of the small country assumption, processing incentives of any form must cause social welfare to fall in this model. This is a well-known result – for the small country no intervention in trade can raise welfare, and this remains true even if the good restricted is a raw material. However, by incorporating foreign ownership of factors of production into this framework, it was shown that export restrictions or other processing incentives may indeed raise the level of welfare for domestic interests through an income transfer effect (although they lower welfare for the world as a whole). But, the result depended on the return to the foreign owners of natural resources used in forestry falling. By extending the model slightly (adding mobile capital) it was shown that this factor return will be of ambiguous sign in general. Also, by extending the model to incorporate many goods and many factors (using the framework of Woodland, 1982), it was shown that few strong predictions about the outcome of policy intervention on outputs or factor returns can be made outside of highly abstract structural models even where the price vector is exogenously determined.

In Chapter 5 we also considered a two country general equilibrium model where an intermediate good and a processed good are simultaneously traded and the price vector is determined endogenously (Section 5.4). This was to enable us to consider the nature of feedback effects in prices, and of foreign intervention in trade. In this model the change in the price vector resulting from policy intervention was shown to be of indeterminate sign in general. Hence the desirable 'feedback' effects that proponents of export restrictions rely on (increasing both the raw material and processed good price simultaneously) are only one theoretical possibility. The result means that no clear-cut predictions about the effect of policy intervention on factor prices, resource allocation, output, or welfare, can be made where it is assumed New Zealand has market power in both lumber and logs without further restrictive assumptions. One possibility that we examined is that New Zealand has market power in the log industry, but not in the lumber industry. In this case the problem collapses to the relatively simple and familiar optimal export tax on logs.

Given the general ambiguity of these theoretical results, counterfactual simulation using computable general equilibrium techniques was considered to be the most appropriate method of analysing the export restriction issue. Given that New Zealand is generally regarded as a relatively small player in world terms, a logical starting point was a single country model with world prices fixed – i.e., an empirical counterpart to the price exogenous models of Chapter 5. The features of the model that we developed were described in Chapter 6. Overall, it falls into the category of the trade-focused, neo-classical CGE, with a special emphasis on the forestry and processing industries (which were separated out and where econometrically estimated variable proportions production functions were used).

The results of simulations using this CGE model (Chapter 7) can be summarised as follows. In the absence of foreign ownership, while it can be shown that export restrictions on logs do result in substantial increases in processing activity, and in the level of processed exports, the welfare costs of such a policy are not insignificant. This is particularly the case with the more restrictive policies, such as a log export ban (where the welfare costs were estimated to be approximately $NZ92 million in the short run, and $NZ48 million in the long run). It was also shown that, if processing is the policy objective, then a given level of output could be achieved at a much lower social cost by other incentives, such as a direct processing subsidy. This confirms another result well-known in trade theory, that if intervention is necessary it should be directly aimed at the policy objective. Trade restrictions are an inherently inefficient means of achieving a processing objective because they distort prices to consumers unnecessarily. Export restrictions on the raw material are particularly costly in welfare terms because they do not even target the industry of interest directly, instead distorting the price of an input for both producers and consumers, resulting in inefficient use of that input.

Regardless of whether export taxes or processing subsidies are used, the short and long run versions of the model used in Chapter 7 indicated that the processing incentives are more effective in terms of the increase in processing that they cause in the long run than in the short run. However, they also result in higher welfare loses in the long run (although the welfare cost per unit of increased output is less). The implication is that a given output target can be achieved at a much lower welfare cost in the long run than in the short run. The increased welfare cost for a given policy in the long run is a reflection of its increased effectiveness, but in the long run resource allocation (with capital mobile) is in fact more efficient than with the same policy in the short run. In order to minimise welfare costs for a given output target then, incentives should be reduced over time.

The one argument that is shown to have some validity is that concerning the degree of foreign ownership. As discussed above, it can be argued that an processing incentives, by lowering the returns to owners of the natural resource, can improve national welfare by transferring income to domestic factors, where those natural resources are owned by foreign interests. Using the theoretical models of Chapter 5 we were able to show that this need not necessarily hold, but the applied experiments of Chapter 7, where the level of foreign ownership and the transfer of income was explicitly modelled, did provide some support for the possibility, provided export restrictions rather than processing incentives were used. By using 'optimal' export taxes to transfer factor incomes, the gain was estimated at approximately $NZ40 and $NZ42 million in the short and long run respectively. If processing subsidies were used the inclusion of foreign ownership actually worsened the results slightly (since returns to the foreign owned factors actually rise). Hence the inclusion of foreign owned factors of production (as with other domestic distortions) can turn the traditional policy prescriptions for the small economy on their head.

In Chapter 8 of the study, we considered trade policy issues from a multi-country perspective, using the GTAP model. There were essentially two reasons for doing this. The first was to reconsider the role of export restrictions in a model which accounted for a wider set of international linkages. This relates directly to both the empirical investigation of the impact of feedback type effects in prices, and to the possibility of using export restrictions as a retaliatory measure against escalating tariffs in forest products importing countries. The second reason was to enable us to directly consider the impact of removing those tariffs on both an individual country and a regional basis.

With respect to the first issue, the results largely come down to size. If New Zealand cannot influence world prices (the mechanism by which the supposed feedback effects take place) then it can have no impact in this sense. The simulations with the GTAP indicated little potential for favourable terms of trade effects with the use of export restrictions. Similarly with respect to retaliation. If New Zealand is not large enough to have any significant impact on prices, then there is little scope for effective retaliation along these lines. Furthermore, the simulations performed in Chapter 8 indicate that the removal of barriers even on a region-wide basis would have little substantial effect on the level of wood processing or welfare within New Zealand. The message is clear: while New Zealand may have a comparative advantage in forestry, it does not have a comparative advantage in wood processing.

Policy Implications

The main policy conclusion that follows from these results is a well-known one. New Zealand is a small country with a comparative advantage in forestry production. It cannot influence world prices to any substantial degree. New Zealand should therefore refrain from intervention in trade as a general principle. The fact that logs are a raw material does not make any exception to this principle. While the use of processing incentives in an attempt to increase the level of processing would benefit some groups (in particular the owners of fixed factors of production in the processing sector), it would also lower the level of overall social welfare. Export restrictions represent an attempt to force resources into an area (wood processing) where they cannot be most efficiently employed.

While the argument that export restrictions can reclaim processing is true in the sense that export restrictions on unprocessed products do result in greater levels of domestic processing, it is deceptive because the implication is that welfare will be improved, which is not in general likely to be the case. There seems to be a perception that increasing the level of processing is an end in itself, and a worthwhile policy objective. Hence, while there is widespread acceptance in New Zealand that restrictions on trade are damaging to the economy, where raw material exports are involved the reasons for this seem to be forgotten. The arguments of the NZOSG cite the fact that with export restrictions the value of exports rise, but this is not the point. The policy forces resources into inefficient uses, and forces consumers to pay distorted prices. In the absence of market power and subsequent offsetting benefits in terms of improved export prices, real income, and hence social welfare, falls. We emphasise again, restricting log exports to favour domestic processing would lower New Zealand welfare under standard assumptions.

If processing incentives are to be used they must be justified on some basis other than increasing overall social welfare. Moreover, export restrictions on logs cannot be justified for a processing objective, since they result in welfare cost that are substantially higher than direct processing subsidies (there also are considerable legal restrictions on the use of export restrictions - in particular export bans - as discussed in Chapter 3). Finally, once again, a given processing objective will require lower levels of subsidisation over longer time horizons, if it is to be achieved at minimum social cost.

The exception to these general policy conclusions is where there is foreign ownership of the forestry resources and the income transfer effect outweighs the deadweight losses. Nevertheless, even in this case such a policy still results in costs for some members of society, and there will be some 'optimal' level of restriction, most likely below the prohibitive level. Such a policy is also of dubious long-term value, since it is essentially one of nationalisation and therefore inconsistent with the goal of attracting foreign investment (although the desirability or otherwise of this is another area of substantial debate).

Limitations of the Study

Although we consider the current study to be the most comprehensive study existing of the impact of processing incentives and other trade policy interventions in forestry on the New Zealand economy, there are a number of important limitations to the study, which should be borne in mind when considering the results presented. Broadly speaking, the limitations fall into two categories, limitations in the methodology and limitations in scope. Most of the limitations have been discussed at some relevant point in the main text, but in this section we briefly outline what we consider to be the main points of note in this respect.

Limitations of the Methodology

With respect to the main methodology used in this study (CGE), we can break down the limitations in terms of the general limitations of the methodology, as discussed in Chapter 6, and the specific limitations of our model. In the first category, perhaps the most important criticism of CGE models is their deterministic nature. Since CGE models are essentially sophisticated theoretical models that have been calibrated to some assumed numerical equilibrium, the results that we obtain are a function of the assumptions made in specifying the model. In this respect it must always be remembered that CGE models are stylised versions of some aspects of a real world economy – versions which hopefully resemble the real economy (or our view of it) in certain key respects, but which nonetheless can never hope to capture all of the intricacies of the real economy under consideration. CGE models are an inductive policy tool. What the results give us is some idea of the likely impact of different policies on different members of society and different industries – and perhaps sensible measures of the magnitude of the sorts of changes which could be expected. The numbers themselves should be interpreted with due caution.

Another common criticism of CGE methodology relates to the use of 'guesstimate' data in setting the parameters of the model. Our model (in Chapter 6) and the GTAP model (in Chapter 8) are also subject to such criticism. In both cases most of the parameters are taken from outside sources. The main exceptions to this are the production elasticities in forestry and wood processing, which were estimated using New Zealand data (Section 6.5). Here too, there are problems, however, since the data used for the forestry sector was not of sufficient quality for us to have much confidence in the results (although the results for the more important processing sector were considerably better). In general we would argue that, given the nature of CGE models and the policy interpretations that can be drawn from them, the issue is not as important as is sometimes made out. Given data limitations and the massive parameter requirements of even relatively small CGE models, such limitations will no doubt persist in the field for some time yet, and so long as they are noted when interpreting the results, and adequate sensitivity analysis is performed, are probably not too damaging.

There are also a number of issues with respect to our use of the GTAP model in particular. The GTAP assumes that logs are used in fixed proportions in the processing industry, in contrast to our model and the empirical evidence. Unfortunately, to eliminate this problem would require re-specification of the GTAP model. We leave this for future research. Also important is the lack of separate hardwood and softwood markets, which may distort some of our results in Chapter 8, an issue discussed further below.

Limitations of Scope

As well as general and specific limitations in methodology, there are also limitations in terms of the scope of our study. One particular point relates back to the general structure of the CGE models used in this study. We have stuck very closely to the traditional, static, neo-classical CGE structure. As discussed in Chapter 6, we believe that there are very good reasons for doing so, not least of which is the fact that the model and its behaviour is relatively easily understood and related back to an underlying structural model. However, it does need to be recognised that the perfectly competitive structure does preclude the analysis certain 'strategic' arguments for intervention. Similarly, the static nature of the model precludes the analysis of certain dynamic arguments for protection. These sorts of arguments and the reasons for their exclusion from the current study were discussed in some detail in Chapter 3. While, as was argued in Chapter 3, we believe that these arguments are in fact of limited applicability for a variety of reasons, it should be remembered that there may be other reasons for encouraging domestic processing that are not adequately dealt with in the framework utilised here. Of course, even if one believes this to be the case, the static costs associated with the policies considered in this study are still of considerable importance – since they provide a base with which to compare estimates of possible gains from other aspects of the problem not considered here.

Other possible economic objectives of processing incentives, such as increasing employment, which were also discussed in Chapter 3, are also not directly considered, nor are non-economic objectives in general (with the exception of carbon emissions in Chapter 7). The exclusion of these issues is not meant to imply that they are irrelevant or unimportant, but arises rather from the need to keep the study to manageable proportions.

Areas for Future Research

There are a number of areas in which the research conducted in this study could be extended. Many of these are concerned with the elimination of some of the limitations of the study in terms of methodology or scope, as were discussed in the preceding section. Thus, for example, we believe there is considerable scope for further econometric estimation of key model parameters for New Zealand, in particular for the forestry industry – if suitable data could be obtained.

There is also considerable scope for work in extending the GTAP model's treatment of the forestry and wood processing industries. In particular we would like to incorporate flexible technology in the use of logs in processing – for the reasons discussed in Chapter 4. It would also be worthwhile to incorporate a more realistic differentiation of hardwood and softwood into the model. This could be accomplished either by defining separate industrial groups for softwood and hardwood forestry and wood products, or by somehow allowing for a separate elasticity of substitution across imports from different regions to be specified. If this were possible we could, for example, set a relatively high elasticity of substitution between New Zealand, Chile and North America, and a relatively low elasticity between these countries and the hardwood producers of East Asia. (At present there is one elasticity governing the choice of source country for imports, for each industrial group).

Another issue that we believe is of substantial importance is that of the presence of environmental externalities. Export restrictions on logs are often put in place under the guise of protecting the environment. Indeed, this is even the case in New Zealand for native forests. By restricting exports of logs, it is hoped that the forests can be protected. However, while this argument may have some validity in the case of an existing natural resource in other countries (although most trade economists would argue that an export ban is an inefficient means of achieving such a goal), in New Zealand the situation is somewhat different. As we discussed in an extension to the basic model in Chapter 7, the New Zealand forestry resource is predominantly exotic plantations, and hence a reduction in forestry output does not represent a reduction in utilisation of an existing resource like a rainforest, but rather a reduction in the creation of a new one (i.e., a reduction in forest plantation). There may well be substantial implications in terms of New Zealand's absorptive capacity for carbon dioxide and other greenhouse gases. Moreover, it is well known that the wood processing sector (in particular pulp and paper) is highly pollutive, and these extra costs should ideally be taken into account. As was shown in Chapter 7, the inclusion of emissions and carbon taxes in the model strengthens the case against export restrictions. There is much more that can be done in this area.

A final area for future research concerns the application of the framework of analysis developed here to other economies. There are many examples of export restrictions on forestry in the region, as discussed in Chapter 3. Moreover, the more general issue of processing incentives is still of considerable interest in many economies, particularly those of developing countries. As discussed in Chapter 4, there is also relatively little empirical research into this area – in particular using general equilibrium analysis. The general modelling framework used here could be adapted to suit other economies and goods where the domestic processing of exported raw materials is considered an issue. We believe there is considerable research potential in this area.

Appendices

A: Derivation of Equations 5.48-5.55 (Chapter 5)

The Marshallian net export function for lumber from the home country is given by:

$$X_2(p_1, p_2, \overline{K}, \overline{L}, \overline{N}) = G_2(p_1, p_2, \overline{K}, \overline{L}, \overline{N}) - D_2(p_2, G(p_1, p_2, \overline{K}, \overline{L}, \overline{N})). \quad \text{(A.1)}$$

Taking the partial derivative of A.1 with respect to p_2 yields:

$$\begin{aligned} X_{22} &= G_{22} - D_{22} - D_{2I}G_2 \\ &= G_{22} - (E_{22} - D_{2I}D_2) - D_{2I}G_2 \\ &= G_{22} - E_{22} + D_{2I}(D_2 - G_2) \\ &= S_{22} + D_{2I}(D_2 - G_2) \\ &= S_{22} - D_{2I}X_2, \end{aligned}$$

where:

$$\begin{aligned} S_{22} &= G_{22} - E_{22}, \\ X_2 &= G_2 - D_2, \end{aligned}$$

by definition, and:

$$D_{22} = E_{22} - D_{2I}D_2,$$

by the Slutsky equation. This demonstrates the derivation of equation 5.48. Equation 50 (for the foreign country) follows similarly.

Taking the partial derivative of equation A.1 with respect to p_1 yields the following expression:

$$\begin{aligned} X_{21} &= G_{21} - D_{2I}G_1 \\ &= S_{21} - D_{2I}G_1 \\ &= S_{21} - D_{21}X_1, \end{aligned}$$

where:

$$G_{21} = S_{21},$$
$$G_1 = X_1,$$

by definition. This is equation 5.49. Again, equation 5.51 follows similarly.

For equations 5.52 to 5.55, we only need to note that, by definition, the following statement is true:

$$G_1(p_1, p_2, \overline{K}, \overline{L}, \overline{N}) = S_1(p_1, p_2, \overline{K}, \overline{L}, \overline{N}) = X_1(p_1, p_2, \overline{K}, \overline{L}, \overline{N}). \qquad \text{(A.2)}$$

Given A.2, it immediately follows that:

$$X_{11} = S_{11} = G_{11},$$
$$X_{12} = S_{12} = G_{12},$$

and similarly for the foreign country.

B: Derivation of the Composite Price and Import Demand Equations

In our model, we choose to aggregate imports and domestic production in the following general manner (equation 6.10):

$$Q = \overline{B} \cdot [\Delta \cdot M^{-\mu} + (1-\Delta) \cdot D^{-\mu}]^{-1/\mu}, \qquad \text{(B.1)}$$

where M is imports, D is domestic production, Q is the Armington composite good, and subscripts indicating goods have been dropped for convenience. This Armington aggregation function is of the constant elasticity of substitution (CES) form, and the procedure for deriving the composite price function is analogous to deriving the unit cost function from a CES production function. The first order conditions for a minimum are:

$$\frac{\partial Q/\partial D}{\partial Q/\partial M} = \frac{(1-\Delta)}{\Delta} \cdot \frac{D^{-(1+\mu)}}{M^{-(1+\mu)}} = \frac{PD}{PM}, \qquad \text{(B.2)}$$

and,

$$Q = \overline{B} \cdot [\Delta \cdot M^{-\mu} + (1-\Delta) \cdot D^{-\mu}]^{-1/\mu} = 1, \tag{B.3}$$

where 1 unit of output is to be produced and it is assumed that $PD > 0$ and $PM > 0$. Raising both sides of equation B.2 to the power of $-1/(1+\mu)$ and rearranging, we obtain the optimal input ratio as:

$$\frac{D}{M} = \left(\frac{\Delta}{1-\Delta}\right)^{-\eta} \cdot \left(\frac{PD}{PM}\right)^{-\eta}, \tag{B.4}$$

where $\eta \equiv 1/(1+\mu)$. Equation 6.14 (the import demand functions) follows immediately by rearrangement of B.4. Also note that by the definition of the domestic use ratio, *D/Q*, we can obtain equation 6.13 by rearranging B.4:

$$D = \left(\frac{\Delta}{1-\Delta}\right)^{-\eta} \cdot \left(\frac{PD}{PM}\right)^{-\eta} \cdot M, \tag{B.5}$$

and dividing through by B.1 to obtain:

$$\overline{B}^{-1} \cdot (PD/(1-\Delta))^{-\eta} / (\Delta \cdot (PM/\Delta)^{\eta \cdot \mu} + (1-\Delta) \cdot (PD/(1-\Delta)^{\eta \cdot \mu})^{-1/\mu}. \tag{B.6}$$

In order to obtain the aggregate price functions, we substitute B.5 into B.3 to obtain:

$$\overline{B} \cdot \left[\Delta \cdot M^{-\mu} + (1-\Delta) \cdot \left(\frac{\Delta \cdot PD}{(1-\Delta) \cdot PM}\right)^{\mu \cdot \eta} \cdot M^{-\mu}\right]^{-\frac{1}{\mu}} = 1. \tag{B.7}$$

Solving for *M* then yields:

$$M = \left(\frac{PM}{\Delta}\right)^{-\eta} \Big/ \overline{B} \cdot \left[\Delta \cdot \left(\frac{PM}{\Delta}\right)^{\mu \cdot \eta} + (1-\Delta) \cdot \left(\frac{PD}{1-\Delta}\right)^{\mu \cdot \eta}\right]^{-\frac{1}{\mu}}. \tag{B.8}$$

Similarly,

$$D = \left(\frac{PD}{1-\Delta}\right)^{-\eta} \Big/ \overline{B} \cdot \left[\Delta \cdot \left(\frac{PM}{\Delta}\right)^{\mu \cdot \eta} + (1-\Delta) \cdot \left(\frac{PD}{1-\Delta}\right)^{\mu \cdot \eta}\right]^{-\frac{1}{\mu}}. \tag{B.9}$$

Equations B.8 and B.9 are the demand functions for imports and domestic production respectively, per unit of the composite. The composite unit price is then given by:

$$\begin{aligned} P &= PM \cdot M + PD \cdot D \\ &= \frac{PM \cdot (PM/\Delta)^{-\eta} + PD \cdot (PD/(1-\Delta))^{-\eta}}{\overline{B} \cdot \left[\Delta \cdot (PM/\Delta)^{\mu \cdot \eta} + (1-\Delta) \cdot (PD/(1-\Delta))^{\mu \cdot \eta}\right]^{-1/\mu}} \\ &= \frac{PM \cdot (PM/\Delta)^{-\eta} + PD \cdot (PD/(1-\Delta))^{-\eta}}{\overline{B} \cdot \left[PM \cdot (PM/\Delta)^{-\eta} + PD \cdot (PD/(1-\Delta))^{-\eta}\right]^{-1/\mu}}. \end{aligned}$$

Since the numerator and the term inside the brackets of the denominator are the same, and $1+1/\mu=1/(1-\eta)$, the price function can be simplified to:

$$\begin{aligned} P &= \overline{B}^{-1} \cdot \left[PM \cdot (PM/\Delta)^{-\eta} + PD \cdot (PD/(1-\Delta))^{-\eta}\right]^{1/(1-\eta)} \\ &= \overline{B}^{-1} \cdot \left[\Delta^{\eta} \cdot PM^{1-\eta} + (1-\Delta)^{\eta} \cdot PD^{1-\eta}\right]^{1/(1-\eta)}, \end{aligned} \tag{B.10}$$

which is equation 6.17 of the model.

C.1: Estimated Cost Shares for New Zealand Wood Processing

Year	Quarter	θ_L	θ_K	θ_W
1984	*Mar*	0.4285	0.2289	0.3427
	Jun	0.4340	0.3011	0.2649
	Sep	0.4065	0.2956	0.2979
	Dec	0.4340	0.2696	0.2964
1985	*Mar*	0.3922	0.2256	0.3823
	Jun	0.4139	0.2684	0.3177
	Sep	0.3994	0.2876	0.3130
	Dec	0.4317	0.2127	0.3556
1986	*Mar*	0.4583	0.1568	0.3850
	Jun	0.4467	0.1692	0.3841
	Sep	0.3839	0.3179	0.2983
	Dec	0.4565	0.0820	0.4616
1987	*Mar*	0.4037	0.2953	0.3010
	Jun	0.4086	0.2377	0.3537
	Sep	0.4352	0.1858	0.3791
	Dec	0.4384	0.3146	0.2471
1988	*Mar*	0.4191	0.1714	0.4095
	Jun	0.3879	0.3654	0.2467
	Sep	0.4120	0.2018	0.3862
	Dec	0.4077	0.3895	0.2027
1989	*Mar*	0.4010	0.3554	0.2435
	Jun	0.3904	0.3886	0.2210
	Sep	0.3886	0.2308	0.3806
	Dec	0.4000	0.4224	0.1776
1990	*Mar*	0.3737	0.3825	0.2438
	Jun	0.3856	0.2743	0.3401
	Sep	0.3959	0.2237	0.3804
	Dec	0.4362	0.2389	0.3249
1991	*Mar*	0.4287	0.1069	0.4644
	Jun	0.4084	0.1938	0.3979
	Sep	0.3961	0.1808	0.4230
	Dec	0.3951	0.2893	0.3156
1992	*Mar*	0.3865	0.2133	0.4002
	Jun	0.3798	0.1504	0.4698
	Sep	0.3752	0.1719	0.4529
	Dec	0.3801	0.2751	0.3448
1993	*Mar*	0.3495	0.2910	0.3596
	Jun	0.3324	0.2855	0.3821
	Sep	0.3407	0.1429	0.5164
	Dec	0.3631	0.2660	0.3709
1994	*Mar*	0.3452	0.2079	0.4469
	Jun	0.3384	0.2069	0.4547
	Sep	0.3401	0.2710	0.3889
	Dec	0.3410	0.3583	0.3008
1995	*Mar*	0.3347	0.2487	0.4166
	Jun	0.3372	0.2533	0.4095
	Sep	0.3330	0.2450	0.4219
	Dec	0.3515	0.2810	0.3675

Source: Estimated from Statistics New Zealand Business Activity Statistics (1995)

C.2: Factor Price Data for New Zealand Wood Processing

Year	Quarter	Log P_L	Log P_K	Log P_W
1984	*Mar*	2.0353	2.5649	6.5554
	Jun	2.0515	2.6391	6.6053
	Sep	2.0498	2.7014	6.7845
	Dec	2.1224	2.6946	6.8512
1985	*Mar*	2.0911	2.9549	6.8669
	Jun	2.1457	3.0910	6.8659
	Sep	2.1526	3.1655	6.8233
	Dec	2.2389	3.0587	6.7890
1986	*Mar*	2.2970	3.0956	6.8804
	Jun	2.3397	2.8792	6.8298
	Sep	2.3224	2.7663	6.7856
	Dec	2.3882	2.8273	6.8189
1987	*Mar*	2.3721	3.2308	6.7946
	Jun	2.3206	2.9549	6.8046
	Sep	2.4351	2.9497	6.7788
	Dec	2.5081	2.8214	6.7946
1988	*Mar*	2.4650	2.7726	6.8211
	Jun	2.4862	2.7663	6.8544
	Sep	2.4603	2.6672	6.8628
	Dec	2.4903	2.6741	6.9460
1989	*Mar*	2.4908	2.5878	6.9137
	Jun	2.4911	2.6027	6.9037
	Sep	2.4762	2.5953	6.9556
	Dec	2.5732	2.6462	7.0255
1990	*Mar*	2.5449	2.6174	7.0211
	Jun	2.5268	2.6247	7.0291
	Sep	2.5037	2.6810	7.0326
	Dec	2.6154	2.5649	7.0379
1991	*Mar*	2.5817	2.4849	7.0282
	Jun	2.5819	2.2925	6.9939
	Sep	2.5686	2.1861	7.0220
	Dec	2.6306	2.0541	7.0519
1992	*Mar*	2.6066	1.9879	7.0767
	Jun	2.5940	1.9169	7.0630
	Sep	2.5897	1.8405	7.1213
	Dec	2.6387	1.9315	7.2049
1993	*Mar*	2.5867	1.9741	7.3467
	Jun	2.5882	1.8563	7.6329
	Sep	2.5997	1.6677	7.7592
	Dec	2.6549	1.6677	7.5616
1994	*Mar*	2.6274	1.6864	7.5267
	Jun	2.5882	1.8083	7.4495
	Sep	2.6086	2.0149	7.3871
	Dec	2.6577	2.2618	7.3389
1995	*Mar*	2.6394	2.2407	7.3421
	Jun	2.6264	2.1972	7.3291
	Sep	2.6207	2.2192	7.2689
	Dec	2.6664	2.1518	7.3052

Source: Statistics New Zealand Business Activity Statistics (1995), New Zealand Forestry Statistics (1995) and Reserve Bank Bulletin (Various Issues)

Bibliography

Adams, D.M. (1987), 'Issues in Trade Modeling: Global versus Regional Models', in P.A. Cardellichio, D.M. Adams, and R.W. Haynes (eds), *Forest Sector and Trade Models: Theory and Applications - Proceedings of an International Symposium*, CINTRAFOR, Seattle.

Adams, D.M. and Haynes, R.W. (1980), 'The 1980 Softwood Timber Assessment Market Model: Structure, Projections, and Policy Simulations', *Forest Science Monograph*, vol. 22, pp. 1-64.

Adams, D.M. and Haynes, R.W. (1986), 'A Spatial Equilibrium Model of US Forest Products Markets for Long Range Projection and Policy Analysis', in M. Kallio, A.E. Andersson, R. Seppala and A. Morgan, (eds), *Systems Analysis in Forestry and Forest Industries*, North-Holland, Amsterdam.

Allen, R.G.D. (1938), *Mathematical Analysis for Economists*, Macmillan Press, London.

Alston, J.M.C., Carter, C.A., Green, R. and Pick, D. (1990), 'Whither Armington Trade Models?', *American Journal of Agricultural Economics*, vol. 72(2), pp. 455-67.

Armington,P.S. (1969), 'A Theory of Demand for Products Distinguished by Place of Production', *IMF Staff Papers*, 16(1), pp. 159-76.

Arndt, S. and Milner, C. (eds) (1995), *The World Economy: Global Trade Policy 1995*, Blackwell Publishers, Oxford.

Arrow, K.J., Chenery, H.B., Minhas, B.S. and Solow, R.M. (1961), 'Capital-Labor Substitution and Economic Efficiency', *Review of Economics and Statistics*, vol. 43(3), pp. 225-50.

Bandara, J.S. (1989), *A Multisectoral General Equilibrium Model of the Sri Lankan Economy with Applications to the Analysis of the Effects of External Shocks*, Unpublished PhD Dissertation, La Trobe University.

Bandara, J.S. (1991), 'Computable General Equilibrium Models for Development Policy Analysis in LDCs', *Journal of Economic Surveys*, vol. 5(1), pp. 3-69.

Barbier, E.B. (1994), 'The Environmental Effects of Trade in the Forestry Sector', in OECD, *The Environmental Effects of Trade*.

Barbier, E.B, Bockstael, N., Burgess, J.C. and Strand, I. (1995), 'The Link Between the Timber Trade and Tropical Deforestation in Indonesia', *World Economy*, vol. 18(3), pp. 411-42.

Barten, A.P. (1969), 'Maximum Likelihood Estimation of a Complete System of Demand Equations', *European Economic Review*, vol. 1(1), pp. 7-73.

Batra, R.N. and Casas, F.R. (1973), 'Intermediate Products and the Pure Theory of International Trade: A Neo-Heckscher-Ohlin Framework', *American Economic Review*, vol. 63(3), pp. 297-311.

Bautista, R.M. (1978), 'Interrelated Products and the Elasticity of Export Supply in Developing Countries', *International Economic Review*, vol. 19, pp. 181-94.

Beladi, H. and Marjit, S. (1992), 'Foreign Capital and Protectionism', *Canadian Journal of Economics*, vol. 15(1), pp. 233-238.

Berndt, E.R. and Christensen, L.R. (1973), 'The Translog Function and the Substitution of Equipment, Structures, and Labor in US Manufacturing 1929-68', *Journal of Econometrics*, vol. 1(1), pp.81-114.

Berndt, E.R. and Wood, D.O. (1975), 'Technology, Prices, and the Derived Demand for Energy', *Review of Economics and Statistics*, vol. 52(3), pp. 259-68

Bhagwati, J.N. (1964), 'The Pure Theory of International Trade: A Survey', *Economic Journal*, vol. 74(293), pp. 1-84.

Bhagwati, J.N. and Brecher, R.A. (1980), 'National Welfare in an Open Economy in the Presence of Foreign Owned Factors of Production', *Journal of International Economics*, vol. 10(1), pp. 103-15.

Bhagwati, J.N., Jones, R.W., Mundell, R.A. and Vanek, J. (eds) (1971), *Trade, Balance of Payments, and Growth: Essays in Honour of Charle P. Kindleberger*, North-Holland, Amsterdam.

Bigsby, H. and Kornai, G. (1987), 'Forest Products Consumption, Production and Trade from an Oceanic Perspective: Adapting the Global Trade Model', in P.A. Cardellichio, D.M. Adams, and R.W. Haynes (eds), *Forest Sector and Trade Models: Theory and Applications - Proceedings of an International Symposium*, CINTRAFOR, Seattle.

Birchfield, R.J. (1993), *Out of the Woods: The Restructuring and Sale of New Zealand's State Forests*, GP Publications, Wellington.

Blatner, K.A. (1987), 'Approaches to Bilateral Trade Modeling', in P.A. Cardellichio, D.M. Adams, and R.W. Haynes (eds), *Forest Sector and Trade Models: Theory and Applications - Proceedings of an International Symposium*, CINTRAFOR, Seattle.

Bocoum, B. and Labys, W.C. (1993), 'Modelling the Economic Impacts of Further Processing: The Case of Zambia and Morocco', *Resources Policy*, vol. 19(4), pp. 247-63.

Boyd, R. and Krutilla, K. (1987), 'The Welfare Impacts of US Trade Restrictions Against the Canadian Softwood Lumber Industry: A Spatial Equilibrium Analysis', *Canadian Journal of Economics*, vol. 20(1), pp. 17-35.

Brecher, R.A. and Diaz-Alexandro, C.F. (1977), 'Tariffs, Foreign Capital and Immiserizing Growth', *Journal of International Economics*, vol. 7, pp. 317-322.

Brecher, R.A. and Findlay, R. (1983), 'Tariffs, Foreign Capital and National Welfare With Sector-Specific Factors', *Journal of International Economics*, vol. 14, pp. 277-288.

Brockmeier, M. (1996), 'A Graphical Exposition of the GTAP Model', GTAP Technical Paper No. 8, Purdue University, Indiana.

Brooke, A., Kendrick, D. and Meeraus, A. (1996), *GAMS: A User's Guide*, Scientific Press, Danvers.

Brooks, D.J. (1987), 'Alternative Approaches to Modeling International Trade: The World Assessment Market Model', in P.A. Cardellichio, D.M. Adams, and R.W. Haynes (eds), *Forest Sector and Trade Models: Theory and Applications - Proceedings of an International Symposium*, CINTRAFOR, Seattle.

Brooks, D.J. (1987b), 'Issues in Trade Modeling: Aggregation of Products', in P.A. Cardellichio, D.M. Adams, and R.W. Haynes (eds), *Forest Sector and Trade Models: Theory and Applications - Proceedings of an International Symposium*, CINTRAFOR, Seattle.

Brown, C. (1997), 'In Depth Country Study - New Zealand', Asia-Pacific Forestry Sector Outlook Study Working Paper No. APFSOS/WP/05, FAO, Paris.

Brown, D., Deardorff, A. and Stern, R.M. (1996), 'Computational Analysis of the Economic Effects of an East Asian Preferential Trading Bloc', *Journal of the Japanese and International Economies*, vol. 10(1), pp. 37-70.

Buongiorno, J. (1986), 'A Model for International Trade in Pulp and Paper', in M. Kallio, A.E. Andersson, R. Seppala and A. Morgan, (eds), *Systems Analysis in Forestry and Forest Industries*, North-Holland, Amsterdam.

Burgess, D.F. (1976), 'The Income Distributional Effects of Processing Incentives: A General Equilibrium Analysis', *Canadian Journal of Economics*, vol. 9(4), pp. 595-612.

Burgess, D.F. (1980), 'Protection, Real Wages, and the Neo-Classical Ambiguity with Inter-Industry Flows', *Journal of Political Economy*, vol. 88(4), pp. 783-802.

Burgess, D.F. (1980b), 'Protection, Real Wages, Real Incomes, and Foreign Ownership', *Canadian Journal of Economics*, vol. 13(4), pp. 594-614.

Burniaux, J., Nicoletti, G. and Olivira-Martins, J. (1992), 'GREEN: A Global Model for Quantifying the Costs of Policies to Curb CO_2 Emissions', Economic Studies No.19, OECD, Paris.

Cardellichio, P.A. and Adams, D.M. (1987), 'U.S. Experience with the Global Trade Model', in P.A. Cardellichio, D.M. Adams, and R.W. Haynes (eds), *Forest Sector and Trade Models: Theory and Applications - Proceedings of an International Symposium*, CINTRAFOR, Seattle.

Cardellichio, P.A., Adams, D.M. and Haynes, R.W. (eds) (1987), *Forest Sector and Trade Models: Theory and Applications - Proceedings of an International Symposium*, CINTRAFOR, Seattle.

Cardellichio, P.A., Youn, Y.C., Adams, D.M., Joo, R.W. and Chmelik, J.T. (1989), 'A Preliminary Analysis of Timber and Timber Products Production, Consumption, Trade, and Prices in the Pacific Rim Until 2000', Working Paper No.22, CINTRAFOR, Seattle.

Casas, F.R. (1972), 'Theory of Intermediate Products, Technical Change and Growth', *Journal of International Economics*, vol. 2(2), pp. 189-200.

Chenery, H. and Srinivasan, T.N. (eds) (1989), *Handbook of Development Economics*, North-Holland, Amsterdam.

Chisholm, A., Moran, A. and Zeitsch, J. (1994), 'The Economic Costs of Stabilising Emissions of Carbon Dioxide in New Zealand', *New Zealand Economic Papers*, vol. 28(1), pp. 1-24.

Christensen, R.L., Jorgensen, D.W. and Lau, L.J. (1971), 'Conjugate Duality and the Transcendental Logarithmic Function', *Econometrica*, vol. 39(4), pp. 255-6.

Clarete, R.L. and Roumasset, J.A. (1986), 'CGE Models and Development Policy Analysis: Problems, Pitfalls, and Challenges', *American Journal of Agricultural Economics*, vol. 68(5), pp. 1212-6.

CNIPS (1983), *Central North Island Planning Study Findings*, Ministry of Works and Development, Wellington.

Corden, W.M. (1966), 'The Structure of the Tariff System and the Effective Protection Rate', *Journal of Political Economy*, vol. 74(3), pp. 231-7.

Corden, W.M. (1969), 'Effective Protection Rates in the General Equilibrium Model: A Geometric Note', *Oxford Economic Papers*, vol. 21(2), pp. 135-41.

Corden, W.M. (1971), 'The Substitution Problem in the Theory of Effective Protection', *Journal of Industrial Economics*, vol. 19(1), pp. 35-57.

Dardis, R. (1967), 'Intermediate Goods and the Gain from Trade', *Review of Economics and Statistics*, vol. 49(4), pp. 502-9.

Dargavel, J. and Tucker, R. (eds) (1992), *Changing Pacific Forests: Historical Perspectives on the Forest Economy of the Pacific Basin*, Forest History Society, Durham.

de Melo, J. (1988), 'Computable General Equilibrium Models for Trade Policy Analysis in Developing Countries: A Survey', *Journal of Policy Modeling*, vol. 10(4), pp. 469-503.

de Melo, J. and Tarr, D. (1992), *A General Equilibrium Analysis of US Trade Policy*, MIT Press, Cambridge.

Deardorff, A.V. and Stern, R.M. (1990), *Computational Analysis of Global Trading Arrangements*, University of Michigan Press, Ann Arbor.

Decaluwé, B. and Martens, A. (1988), 'CGE Modeling and Developing Economies: A Concise Empirical Survey of 73 Applications to 26 Countries', *Journal of Policy Modeling*, vol. 10(4), pp. 529-568.

Dervis, K., de Melo, J. and Robinson, S. (1982), *General Equilibrium Models for Development Policy*, Cambridge University Press, New York.

Devarajan, S. and Lewis, J.D. (1990), 'Policy Lessons from Trade-Focused, Two-Sector Models', *Journal of Policy Modeling*, vol. 12(4), pp. 625-657.

Devarajan, S. and Rodrik, D. (1989), 'Trade Liberalisation in Developing Countries: Do Imperfect Competition and Scale Economies Matter?', *American Economic Review*, vol. 79(2), pp. 283-7.

Devarajan, S., Lewis, J.D. and Robinson, S. (1986), 'A Bibliography of Computable General Equilibrium Models Applied to Developing Countries' Harvard Institute for International Development Discussion Paper No.22.

Dixit, A. and Norman, V. (1980), *Theory of International Trade: A Dual General Equilibrium Approach*, Cambridge University Press, Cambridge.

Dixon, P.B., Paramenter, B.R., Sutton, J. and Vincent, D.P. (1982), *ORANI: A Muiti-sectoral Model of the Australian Economy*, North-Holland, Amsterdam

Edgar, M.J., Lee, D. and Quinn, B.P. (1992), *New Zealand Forest Industries Strategy Study*, New Zealand Forest Industries Council, Wellington.

Edlin, B. (1993), 'No Easy Answers in Log Export Debate', *New Zealand Forest Industries*, March, p. 11.

Edlin, B. (1995), 'Apple Pie Takeaways', *New Zealand Forest Industries*, June, pp. 42-3.

Elliot, D.A. and Levack, H.H. (1981), 'New Zealand Plantation Resource - Areas, Locations, Quantities', Paper to Forestry Conference.

Espinosa, J.A. and Smith, V.K. (1995), 'Measuring the Environmental Consequences of Trade Policy: A Non-market CGE Approach', *American Journal of Agricultural Economics*, vol.77(3), pp.772-77.

Familton, A.K. (1969), 'The 1969 National Planning Model', Paper to Forestry Development Conference.

FAO (1988), 'Trade in Forest Products: A Study of the Barriers Faced by the Developing Countries' Forestry Paper No.83.

FAO (1993), *Yearbook of Forest Products.*

FAO (1995), *1945-1993...2010 Forestry Statistics Today for Tomorrow.*

Fleming, G. (1990), *Economic Thought and Policy Advice in New Zealand: Economists and the Agricultural Sector*, Unpublished PhD Dissertation, University of Auckland.

Flora, D. and McGinnis, W. (1989), 'Embargoes on and off: Some Effects of Ending the Export Ban on Federal Logs and Halting Exports of State-Owned Logs' Working Paper No.19, CINTRAFOR, Seattle.

Fuss, M. and McFadden, D. (1978), *Production Economics: A Dual Approach to Theory and Application*, North-Holland, Amsterdam.

Gillis, M. (1988), 'Indonesia: Public Policies, Resource Management, and the Tropical Forest' in R. Repetto and M. Gillis (eds) *Public Policies and the Misuse of Forest Resources*, Cambridge University Press, New York.

Golub, S.S. and Finger, J.M. (1979), 'The Processing of Primary Commodities: Effects of Developed Country Tariff Escalation and Developing Country Export Taxes', *Journal of Political Economy*, vol. 187(3), pp.559-77.

Grant, H.E., Lowen, K. and Smith, A.W. (1978), *A Core Model for Exploring Forestry Sector Strategies*, Ministry of Works and Development, Wellington.

Grant, H.E., Smith, A.W., Bartlett, A.M., Cavana, R.Y. and Lowen, K.D. (1979), *A User's Guide to the MWD Forestry Model*, Ministry of Works and Development, Wellington.

Greenaway, D. and Milner, C. (1993), *Trade and Industrial Policy in Developing Countries: A Manual of Policy Analysis*, MacMillan Press, London.

Hanoch, G. (1971), 'CRESH Production Functions', *Econometrica*, vol. 39(5), pp. 695-712.

Harrison, G. and Kimbell, L. (1985), 'Economic Interdependence in the Pacific Basin: A General Equilibrium Approach', in Piggott, J. and Whalley, J. (eds), *New Developments in Applied General Equilibrium Analysis*, Cambridge University Press, New York.

Harrison, J. and Pearson, K. (1996), *GEMPACK User Documentation Release 5.2 (Vol 1-3)*, Impact Project and KPSOFT, Melbourne.

Haynes, R. (1976), *Price Impacts of Log Export Restrictions under Alternative Assumptions*, US Department of Agriculture, Washington DC.

Haynes, R. (1977), 'A Derived Demand Approach to Estimating the Linkage between Stumpage and Lumber Markets', *Forest Science*, vol. 23, pp. 281-8.

Hazari, B.R. and Pattanaik, P.K. (1980), 'Some Welfare Propositions in a Three Commodity, Three Factor Model of Trade in the Presence of Foreign-Owned Factors of Production', *Greek Economic Review*, vol. 2(1), pp. 12-3.

Helpman, E. and Krugman, P.R. (1989), *Trade Policy and Market Structure*, MIT Press, Cambridge.

Henderson, D. (1996), *Economic Reform: New Zealand in an International Perspective*, New Zealand Business Roundtable, Wellington.

Hertel, T.W. (ed) (1997), *Global Trade Analysis: Modeling and Applications*, Cambridge University Press, New York.

Horton, M. (1995), *Clearcut: Forestry in New Zealand*, Campaign Against Foreign Control of Aotearoa, Christchurch.

Hosking, M.R. (1972), *The 1972 National Planning Model*, FCD Report, Rotorua.

Hunter, L.A.J. (1987), 'Economic Modeling On and In Forestry', in A.G.D. Whyte (ed), *8th NZASIA Conference on Asia Studies Forestry Papers*, School of Forestry, University of Canterbury.

Intrilligator, M. and Kendrick, D. (eds) (1974), *Frontiers in Quantitative Economics*, North-Holland, Amsterdam.

Johansen, L. (1960), *A Multisectoral Study of Economic Growth*, North-Holland, Amsterdam.

Johansson, P.O. and Lofgren, K.G. (1985), *The Economics of Forestry and Natural Resources*, Basil Blackwell, Oxford.

Johnson, S.R. (1986), 'Doable General Equilibrium Models: Discussion', *American Journal of Agricultural Economics*, vol. 68(5), pp. 1217-8.

Jomini, P., Zeitsch, J.F., McDougall, R. Welsh, A., Brown, S., Hambley, J. and Kelly, J. (1991), *SALTER: A General Equilibrium Model of the World Economy*, Industry Commission, Canberra.

Jones, K.A. (1994), *Export Restraint and the New Protectionism: The Political Economy of Discriminatory Trade Restrictions*, University of Michigan Press, Ann Arbor.

Jones, R.W. (1971), 'Effective Protection and Substitution' *Journal of Industrial Economics*, vol. 19(1), pp. 59-81.

Jones, R.W. and Kenen, P.B. (eds) (1984), *Handbook of International Economics Vol.I*, North-Holland, Amsterdam.

Jones, R.W. and Spencer, B. (1989), 'Raw Materials, Processing Activities, and Protectionism', *Canadian Journal of Economics*, vol. 22(3), pp. 469-86.

Kallio, M., Andersson, A.E., Seppala, R. and Morgan, A. (eds) (1986), *Systems Analysis in Forestry and Forest Industries*, North-Holland, Amsterdam.

Kallio, M., Dykstra, D.P. and Binkley, C.S. (1987), *The Global Forest Sector: An Analytical Perspective*, John Wiley & Sons, Chichester.

Kemp, M.C. (1969), *The Pure Theory of International Trade and Investment*, Prentice Hall, New Jersey.

Kemp, M.C. and Long, N.V. (1984), 'The Role of Natural Resources in Trade Models', in R.W. Jones and P.B. Kenen (eds), *Handbook of International Economics Vol.I*, North-Holland, Amsterdam.

Keppler, H. (1985), 'Commodity Export Taxes as a Means of Promoting Internal Processing Industries: A General Equilibrium Model', in J. Weinblatt (ed), *The Economics of Export Restrictions: Free Access to Commodity Markets and the NIEO*, Westview Press, Boulder.

Keyzer, M.A. and van Veen, W.C.M. (1994), 'Food Policy Simulations for India: The Fifth Five Year Plan Period 1989-1993', in J. Mercenier and T.N. Srinivasan (eds), *Applied General Equilibrium and Economic Development*, University of Michigan Press, Ann Arbor.

Kininmonth, J.A. (1997), *A History of Forestry Research in New Zealand: Commemorating 50 Years of Research at the Forest Research Institute (FRI)*, New Zealand Forest Research Institute, Rotorua.

Kmenta, J. and Gilbert, R.F. (1968), 'Small Sample Properties of Alternative Estimators of Seemingly Unrelated Regressions', *Journal of the American Statistical Association*, vol. 63(324), pp. 1180-200.

Kohli, U. (1991), *Technology, Duality and Foreign Trade: The GNP Function Approach to Modeling Imports and Exports*, University of Michigan Press, Ann Arbor.

Krugman, P.R. (1987), *Strategic Trade Policy and the New International Economics*, MIT Press, Cambridge.

Kumar, R.C. (1988), 'On Optimal Domestic Processing of Exhaustible Natural Resource Exports', *Journal of Environmental Economics and Management*, vol. 15(3), pp. 341-54.

Kuska, E. (1973), *Maxima, Minima and Comparative Statics*, Weidenfeld and Nicolson, London.

Lancaster, K. (1968), *Mathematical Economics*, MacMillan, New York.

Lee, H., Roland-Holst, D. and van de Mensbrugghe, D. (1997), 'APEC Trade Liberalisation and Structural Adjustment: Policy Assessment' APEC Study Centre Discusion Paper No.13, Nagoya University.

Le Heron, R. (1997), 'The Fibre of Forestry: A Perspective on Structural Changes and Challenges Relating to New Zealand's Forestry Scene' Paper Presented to the Australia and New Zealand Institute of Foresters Conference, Canberra 21-24 April 1997.

Leslie, A.J. (1986), 'Forestry Sector Analysis and Planning', New Zealand Forestry Council Working Paper No.9.

Levack, H.H. (1979), *The 1979 National Forestry Planning Model*, Unpublished manuscript.

Lewis, J.D., Robinson, S. and Wang, Z. (1995), 'Beyond the Uruguay Round: The Implications of an Asian Free Trade Area', World Bank Policy Research Working Paper #1467.

Lin, C-I. (1993), *Processing and Exporting Raw Materials: An Analysis of the Indonesian Log and Plywood Industries*, Unpublished PhD Dissertation, University of Michigan.

Lindsay, H. (1989), 'The Indonesian Log Export Ban: An Estimation of Foregone Export Earnings', *Bulletin of Indonesian Economic Studies*, vol. 25(2), pp. 111-23.

Lippke, B. and Perez-Garcia, J. (1992), 'The Economic Impact of a Log Export Tax: Who Gains and Who Loses', Special Paper SP02, CINTRAFOR, Seattle.

Lippke, H. (1994), 'The Economic Effects of the Forest Resources Conservation and Shortage Relief Act on Timber Sales' Working Paper No.25A, CINTRAFOR, Seattle.

Löfgren, K.G. and Johansson, P.O. (1983), *Forest Economics and the Economics of Natural Resources*, Swedish University of Agricultural Sciences.

Lönnstedt, L. (1986), 'A Dynamic Simulation Model for the Forest Sector with an Illustration for Sweden', in M. Kallio, A.E. Andersson, R. Seppala and A. Morgan, (eds), *Systems Analysis in Forestry and Forest Industries*, North-Holland, Amsterdam.

Loo, T. and Tower, E. (1990), 'Agricultural Liberalization, Welfare, Revenue and Nutrition in Developing Countries', in. I. Goldin and O. Knudsen (eds), *Agricultural Trade Liberalization: Implications for Developing Countries*, OECD, Paris and World Bank, Washington DC.

Lowen, K. and Philpott, B. (1980), 'Interfacing Industry Level and Economy-Wide Models for Planning - An Example From Forestry', PEP Discussion Paper No.17, Victoria University of Wellington.

Marks, R.E., Swan, P.L, McLennan, P., Schodde, R., Dixon, P.B., and Johnson, D.T. (1991), 'The Costs of Australian Carbon Dioxide Abatement', *Energy Journal*, vol. 12(2), pp. 135-52.

Martin, W., Petri, P.A. and Yanagishima, K. (1994), 'Charting the Pacific: An Empirical Assessment of Integration Initiatives', *International Trade Journal*, vol. 8(4), pp. 447-82.

Maughan, C.W. (1986), 'The Market for Sawn Timber and Panel Products (Exotic Softwoods) in New Zealand', Centre for Agricultural Policy Studies, Massey University.

McDougall, R.A. (ed) (1997), 'Global Trade, Assistance, and Protection: The GTAP 3 Database', Center for Global Trade Analysis, Purdue University.

McDougall, R.A. (ed) (1998), 'Global Trade, Assistance, and Protection: The GTAP 4 Database', Center for Global Trade Analysis, Purdue University.

McFadden, D. (1978), 'Estimation Techniques for the Elasticity of Substitution and Other Production Parameters', M. Fuss and D. McFadden, *Production Economics: A Dual Approach to Theory and Application*, North-Holland, Amsterdam.

McGovern, E. (1995), *International Trade Regulation*, Globefield Press, Exeter.

McKenzie, L. (1951), 'Ideal Output and the Interdependence of Firms', *Economic Journal*, vol. 61(244), pp. 785-803.

Mercenier, J. and Srinivasan, T.N. (eds.) (1994), *Applied General Equilibrium and Economic Development*, University of Michigan Press, Ann Arbor.

Ministry of Foreign Affairs and Trade (1994), *Trading Ahead: The GATT Uruguay Round: Results for New Zealand.*

Ministry of Forestry (1988), *The Forestry Sector in New Zealand.*

Ministry of Forestry (1988b), *The Forest Industries: New Horizons (Seminar Proceedings).*

Ministry of Forestry (1989), *Supporting New Zealand Forestry.*

Ministry of Forestry (1990), *Post Election Briefing.*

Ministry of Forestry (1992), *Forestry: Investment Opportunities in the New Zealand Forest Industry.*

Ministry of Forestry (1994), *Exporting Lumber and Remanufactured Products to Australia and the United States.*

Ministry of Forestry (1996), *New Zealand Forestry Statistics 1995.*

Moffett, J.L. and Waggener, T.R. (1992), 'The Development of the Japanese Wood Trade: Historical Perspective and Current Trends', Working Paper No.38, CINTRAFOR, Seattle.

New Zealand Forest Owners Association (1996), *Forestry Facts and Figures.*

Nilsson, S. (1987), 'A Critical Review of the GTM: A Swedish Viewpoint', in P.A. Cardellichio, D.M. Adams, and R.W. Haynes (eds), *Forest Sector and Trade Models: Theory and Applications - Proceedings of an International Symposium*, CINTRAFOR, Seattle.

OECD (1994), *The Environmental Effects of Trade.*

Panagariya, A. (1994), 'East Asia and the New Regionalism in World Trade', *World Economy*, vol. 17(6), pp. 817-39.

PECC (1995), *Milestones in APEC Liberalisation: A Map of Market Opening Measures by APEC Economies: A Report by the Pacific Economic Cooperation Council for APEC.*

Percy, M. (1986), *Forest Management and Economic Growth in British Columbia*, Economic Council of Canada, Ottawa.

Percy, M. and Constantino, L. (1987), 'A Policy Simulation Model for the Forest Sectors of British Columbia and Canada', in P.A. Cardellichio, D.M. Adams, and R.W. Haynes (eds), *Forest Sector and Trade Models: Theory and Applications - Proceedings of an International Symposium*, CINTRAFOR, Seattle.

Perez-Garcia, J.M. (1991), 'An Assessment of the Impacts of Recent Environmental and Trade Restrictions on Timber Harvest and Exports', Working Paper No.33, CINTRAFOR, Seattle.

Perez-Garcia, J.M. (1993), 'Global Forestry Impacts of Reducing Softwood Supplies from North America', Working Paper 43, CINTRAFOR, Seattle.

Perez-Garcia, J.M., Lippke Fretwell, H., Lippke, B. and Yu, X. (1994), 'The Impact on Domestic and Global Markets of a Pacific Northwest Log Export Ban or Tax', Working Paper 47, CINTRAFOR, Seattle.

Perroni, C. and Wigle, R. (1997), 'Environmental Policy Modeling', in T.W. Hertel (ed), *Global Trade Analysis: Modeling and Applications*, Cambridge University Press, New York.

Philpott, B.P. and Elley, V.C. (1974), 'A Planning Model of the New Zealand Economy with Revised Data', PEP Occasional Paper No.24, Victoria University of Wellington.

Philpott, B.P. and Elley, V.C. (1974b), 'The Victoria Planning Model: A New Version Incorporating Alternative Capital-Labour Ratios and Import Substitution', PEP Occasional Paper No.25, Victoria University of Wellington.

Ramaswami, V.K. and Srinivasen, T.N. (1971), 'Tariff Structure and Resource Allocation in the Presence of Substitution', in J.N. Bhagwati, R.W. Jones, R.A. Mundell and J. Vanek (eds), *Trade, Balance of Payments, and Growth: Essays in Honour of Charle P. Kindleberger*, North-Holland, Amsterdam.

Ray, A. (1975), 'Traded and Non-Traded Intermediate Inputs and Some Aspects of the Pure Theory of International Trade', *Quarterly Journal of Economics*, vol. 89(2), pp. 331-40.

Repetto, R. and Gillis, M. (eds) (1988), *Public Policies and the Misuse of Forest Resources*, Cambridge University Press, New York.

Riedel, J. (1976), 'Intermediate Products and the Pure Theory of International Trade: A Generalization of the Pure Intermediate Good Case', *American Economic Review*, vol. 66(3), pp. 441-7.

Robinson, S. (1989), 'Multisectoral Models', in H. Chenery and T.N. Srinivasan (eds), *Handbook of Development Economics*, North-Holland, Amsterdam.

Roche, M.M. (1990), *History of Forestry*, New Zealand Forestry Corporation and GP Books, Wellington.

Roche, M.M. (1992), 'Privatising the Exotic Forest Estate: The New Zealand Experience', in J. Dargavel and R. Tucker (eds), *Changing Pacific Forests: Historical Perspectives on the Forest Economy of the Pacific Basin*, Forest History Society, Durham.

Roche, M.M. and Reeves, L. (1987), 'Westland and the Timber Export Regulations 1918-1928', Working Paper 87/4, New Zealand Forest Service, Rotorua.

Roemer, M. (1979), 'Resource-Based Industrialization in the Developing Countries: A Survey', *Journal of Development Economics*, vol. 6(2), pp. 163-202.

Rom, M. (1985), 'Analysis of the GATT Provisions', in J. Weinblatt (ed), *The Economics of Export Restrictions: Free Access to Commodity Markets and the NIEO*, Westview Press, Boulder.

Rom, M. (1985b), 'Export Controls: An Institutional and Historical Perspective', in J. Weinblatt (ed), *The Economics of Export Restrictions: Free Access to Commodity Markets and the NIEO*, Westview Press, Boulder.

Samuelson, P. (1953), 'Prices of Factors and Goods in General Equilibrium', *Review of Economic Studies*, vol. 21(54), pp. 1-20.

Sanyal, K.K. and Jones, R.W. (1982), 'The Theory of Trade in Middle Products', *American Economic Review*, vol. 72(1), pp. 16-31.

Sato, K. (1967), 'A Two-Level Constant Elasticity of Substitution Production Function', *Review of Economic Studies*, vol. 34(1), pp. 201-18.

Sazanami, Y., Urata, S. and Kawai, H. (1995), *Measuring the Costs of Protection in Japan*, Institute for International Economics, Washington DC.

Scarf, H.E. and Shoven, J.B. (eds) (1984), *Applied General Equilibrium Analysis*, Cambridge University Press, New York.

Schmitt, G.J. (1972), *Sale of Interests in State Forests (Report to the Honourable Minister of Forests)*, School of Management Studies, University of Waikato.

Schweinberger, A.G. (1975), 'Pure Traded Intermediate Products and the Heckscher-Ohlin Theorem', *American Economic Review*, vol. 65(4), pp. 634-43.

Sedjo, R.A. (1983), *The Comparative Economics of Plantation Forestry: A Global Assessment*, Resources for the Future, Washington DC.

Sedjo, R.A. and Wiseman, A.C. (1983), 'The Effectiveness of an Export Restriction on Logs', *American Journal of Agricultural Economics*, vol. 65(1), pp. 113-6.

Shephard, R.W. (1970), *The Theory of Cost and Production Functions*, Princeton, New Jersey.

Shoven, J.B. and Whalley, J. (1974), 'On the Computation of Competitive Equilibrium in International Markets with Tariffs', *Journal of International Economics*, vol. 4(4), pp. 341-54.

Shoven, J.B. and Whalley, J. (1984), 'Applied General Equilibrium Models of Taxation and International Trade: An Introduction and Survey', *Journal of Economic Literature*, vol. 22(3), pp. 1007-51.

Shughart, W. (1990), *The Organisation of Industry*, Richard Irwin, Homewood.

Sidabutar, H. (1988), *An Investigation of the Impacts of Domestic Log Processing and Log Export Restrictions on Indonesia's Export Earnings from Logs, Lumber, Plywood*, Unpublished PhD Dissertation, University of Washington.

Silverstone, B., Bollard, A. and Lattimore, R. (eds) (1996), *A Study of Economic Reform: The Case of New Zealand*, North-Holland, Amsterdam.

Srinivasan, T.N. (1983), 'International Factor Movements, Commodity Trade and Commercial Policy in a Specific Factor Model', *Journal of International Economics*, vol. 14(3), pp. 289-312.

Srinivasan, T.N. and Whalley, J. (eds) (1986), *General Equilibrium Trade Policy Modeling*, MIT Press, Cambridge.

Statistics New Zealand (1991), *Four Firm Concentration Ratios by Industry.*

Statistics New Zealand (1993), *Yearbook of New Zealand.*

Statistics New Zealand (1996), *Key Statistics.*

Suzuki, K. (1978), 'The Welfare Effects of an Export Tax Levied on an Intermediate Good', *Quarterly Journal of Economics*, vol. 92(1), pp. 55-69.

Svensson, L.E.O. (1984), 'Factor Trade and Goods Trade', *Journal of International Economics*, vol. 16(4), pp. 365-78.

Swarz, A. (1992), 'Timber is the Test: Forestry Controls Dampen Export Earnings', *Far Eastern Economic Review*, 23 July, p. 36.

Swier, N.P.C. (1993), *Foreign Investment and Restructuring in the New Zealand Forest Industry*, Unpublished Master's Thesis, University of Auckland.

Takeuchi, K. (1983), 'Mechanical Processing of Tropical Hardwood in Developing Countries: Issues and Prospects for the Plywood Industry's Development in the Asia-Pacific Region', in World Bank (1983), *Case Studies on Industrial Processing of Primary Products Vol.1.*

Tawada, M. (1981), *International Trade with a Replenishable Resource: The Steady State Analysis*, Kobe University of Commerce.

Tawada, M. (1982), 'A Note on International Trade with a Renewable Resource', *International Economic Review*, vol. 23(1), pp. 157-63.

Taylor, L. and Black, S.L. (1974), 'Practical General Equilibrium Estimation of Resource Pulls Under Trade Liberalization', *Journal of International Economics*, vol. 4(1), pp. 341-54.

UNIDO (1983), *First World-wide Study of the Wood and Wood Processing Industries*, Sectoral Studies Series No.2.

Uzawa, H. (1962), 'Production Functions with Constant Elasticities of Substitution', *Review of Economic Studies*, vol. 29(81), pp. 291-9.

Valuation New Zealand (1996), *Rural Property Sales Statistics.*

Vanek, J. (1963), 'Variable Factor Proportions and Inter-Industry Flows in the Theory of International Trade', *Quarterly Journal of Economics*, vol. 77(1), pp. 129-42.

Varian, H. (1984), *Microeconomic Analysis*, Norton, New York.

Vernon, J. and Graham, D. (1971), 'Profitability of Monopolization by Vertical Integration', *Journal of Political Economy*, vol. 79(4), pp. 924-5.

Vousden, N. (1974), 'International Trade and Exhaustible Resources: A Theoretical Model', *International Economic Review*, vol. 15(1), pp. 149-67.

Wall, D. (1980), 'Industrial Processing of Natural Resources', *World Development*, vol. 8(4), pp. 303-16.

Weeks, W.J. (1987), 'Using the IIASA Global Trade Model to Examine the Location Effects of Ocean Shipping on Developing Country Timber Producers', in P.A. Cardellichio, D.M. Adams, and R.W. Haynes (eds), *Forest Sector and Trade Models: Theory and Applications - Proceedings of an International Symposium*, CINTRAFOR, Seattle.

Weinblatt, J. (ed) (1985), *The Economics of Export Restrictions: Free Access to Commodity Markets and the NIEO*, Westview Press, Boulder.

Weiner, A.A. (1973), 'Export of Forest Products: Would Cutting Off Log Exports Lower Prices of Wood Products?', *Journal of Forestry*, pp. 215-6.

Whalley, J. (1985), 'Hidden Challenges in Recent Applied General Equilibrium Exercises', in Piggott, J. and Whalley, J. (eds), *New Developments in Applied General Equilibrium Analysis*, Cambridge University Press, New York.

Whyte, A.G.D. (1989), 'An Appropriate Framework for Modelling the New Zealand Forest Sector', in A.G.D. Whyte (ed), *8th NZASIA Conference on Asia Studies Forestry Papers*, School of Forestry, University of Canterbury.

Whyte, A.G.D. (ed) (1989b), *8th NZASIA Conference on Asia Studies Forestry Papers*, School of Forestry, University of Canterbury.

Wije-wardana, D. (1989), 'Current Forest Products Trade Patterns in Asia and Pacific Rim Countries', in A.G.D. Whyte (ed), *8th NZASIA Conference on Asia Studies Forestry Papers*, School of Forestry, University of Canterbury.

Wiseman, A.C. and Sedjo, R.A (1981), 'Effects of an Export Embargo on Related Goods: Logs and Lumber', *American Journal of Agricultural Economics*, vol. 63(3), pp. 423-9.

Wonnacott, P. (1995), 'Merchandise Trade in the APEC Region: Is There Scope for Liberalisation on an MFN Basis?', in S. Arndt and C. Milner (eds), *The World Economy: Global Trade Policy 1995*, Blackwell Publishers, Oxford.

Woodland, A.D. (1977), 'Joint Outputs, Intermediate Inputs and International Trade Theory', *International Economic Review*, vol. 18(3), pp. 517-33.

Woodland, A.D. (1982), *International Trade and Resource Allocation*, North-Holland, Amsterdam.

Woodland, A.D. (1983), 'Stability, Capital Mobility, and Trade', *International Economic Review*, vol. 24(2), pp. 475-83.

World Bank (1983), *Case Studies on Industrial Processing of Primary Products Vol.1*.

Yamazawa, I. (1992), 'On Pacific Economic Integration', *Economic Journal*, vol. 102(415), pp. 1519-29.

Yamazawa, I. (1994), 'Asia Pacific Economic Community: New Paradigms and Challenges', *Journal of Asian Economics*, vol. 5(3), pp. 301-12.

Yamazawa, I. (1996), 'APEC's New Development and Its Implications for Non-member Developing Countries', *Developing Economies*, vol. 37(2), pp. 113-37.

Yeats, A.J. (1981), 'The Influence of Trade and Commercial Barriers on Industrial Processing of Natural Resources', *World Development*, vol. 9(5), pp. 485-94.

Young, L.M. and Huff, K.M. (1997), 'Free Trade in the Pacific Rim: On What Basis?', in T.W. Hertel (ed), *Global Trade Analysis: Modeling and Applications*, Cambridge University Press, New York.

Author Index

Subject Index